SUSPECT FAMILIES

Research in Migration and Ethnic Relations Series

Series Editor:
Maykel Verkuyten, ERCOMER
Utrecht University

The Research in Migration and Ethnic Relations series has been at the forefront of research in the field for ten years. The series has built an international reputation for cutting edge theoretical work, for comparative research especially on Europe and for nationally-based studies with broader relevance to international issues. Published in association with the European Research Centre on Migration and Ethnic Relations (ERCOMER), Utrecht University, it draws contributions from the best international scholars in the field, offering an interdisciplinary perspective on some of the key issues of the contemporary world.

Also in series

Intercultural Education in the European Context
Theories, Experiences, Challenges
Edited by Marco Catarci and Massimiliano Fiorucci
ISBN 978 1 4724 5162 0

Diasporas and Homeland Conflicts
A Comparative Perspective
Bahar Baser
ISBN 978 1 4724 2562 1

Full series list at back of book

EUROPEAN RESEARCH CENTRE ON MIGRATION & ETHNIC RELATIONS

Suspect Families

DNA Analysis, Family Reunification and Immigration Policies

Edited by

TORSTEN HEINEMANN
University of Hamburg, Germany

ILPO HELÉN
University of Eastern Finland, Finland

THOMAS LEMKE
Goethe University Frankfurt, Germany

URSULA NAUE
University of Vienna, Austria

MARTIN G. WEISS
University of Klagenfurt, Austria

Routledge
Taylor & Francis Group
LONDON AND NEW YORK

First published 2015 by Ashgate Publishing

Published 2016 by Routledge
2 Park Square, Milton Park, Abingdon, Oxon OX14 4RN
711 Third Avenue, New York, NY 10017, USA

Routledge is an imprint of the Taylor & Francis Group, an informa business

British Library Cataloguing in Publication Data
A catalogue record for this book is available from the British Library

The Library of Congress has cataloged the printed edition as follows:
Heinemann, Torsten.
Suspect families : DNA analysis, family reunification and immigration policies / by Torsten Heinemann, Ilpo Helén, Thomas Lemke, Ursula Naue and Martin G. Weiss.
pages cm. -- (Research in migration and ethnic relations series)
Includes bibliographical references and index.
ISBN 978-1-4724-2424-2 (hardback)
1. Europe--Emigration and immigration--Government policy.
2. Family reunification--Europe. 3. DNA--Analysis.
I. Title.
JV7590.H45 2015
325.4--dc23

2014030473

ISBN 9781472424242 (hbk)

Contents

Notes on Contributors		*vii*
Acknowledgements		*ix*
1	Constellations, Complexities and Challenges of Researching DNA Analysis for Family Reunification: An Introduction *Torsten Heinemann, Ilpo Helén, Thomas Lemke, Ursula Naue and Martin G. Weiss*	1
2	Germany: The Geneticisation of the Family *Torsten Heinemann and Thomas Lemke*	13
3	Finland: Securing Human Rights, Suspecting Fraud *Anna-Maria Tapaninen and Ilpo Helén*	33
4	Austria: DNA Profiling as a Lie Detector *Kevin Hall and Ursula Naue*	55
5	Ethical Aspects of DNA Testing for Family Reunification *Martin G. Weiss*	79
6	Governing DNA Analysis for Family Reunification: A Comparative Perspective *Torsten Heinemann, Ilpo Helén, Thomas Lemke, Ursula Naue and Martin G. Weiss*	93
References		*111*
Index		*127*

Notes on Contributors

Kevin Hall is a research assistant at the Institute for Cultural Anthropology and European Ethnology at the Goethe University Frankfurt in Germany. He has a background in biochemistry, sociology and philosophy. His research draws on STS and Surveillance Studies to explore the social implications of scientific innovations in the areas of health and biomedicine. In his dissertation project he is investigating the multiple translations and transformations employed by the influenza surveillance system in Germany to aggregate data from disparate sources and to visualise influenza activity.

Torsten Heinemann is Professor of Sociology with a focus on Social Problems and Social Control at the Faculty of Economics and Social Sciences at the University of Hamburg, and Associate Senior Research Fellow in the Biotechnologies, Nature and Society Research Group in the Faculty of Social Sciences at the Goethe University Frankfurt in Germany. His research interests are in social theory, cultural sociology and social studies of science and technology with a special focus on recent developments in the neurosciences and genetic testing. Recent publications include *Populäre Wissenschaft: Hirnforschung zwischen Labor und Talkshow* (Wallstein 2012); and *Risky Profiles: Societal Dimensions of Forensic Uses of DNA Profiling Technologies* (co-edited with Thomas Lemke and Barbara Prainsack), Special Issue of *New Genetics and Society*, 2012.

Ilpo Helén is Professor of Sociology in the Department of Social Sciences at the University of Eastern Finland, Joensuu, and a docent of sociology at the University of Helsinki in Finland. His current research focuses on the biopolitics and political economy of contemporary biomedical technologies, particularly on (post)genomics and neurosciences.

Thomas Lemke is Professor of Sociology with a focus on Biotechnologies, Nature and Society in the Faculty of Social Sciences at the Goethe University Frankfurt in Germany. His research interests cover social and political theory, biopolitics, and social studies of genetic and reproductive technologies. Recent publications include *Perspectives on Genetic Discrimination* (Routledge 2013); *Governmentality: Current Issues and Future Challenges* (co-edited with Ulrich Bröckling and Susanne Krasmann; Routledge 2011); *Biopolitics: An Advanced Introduction* (New York University Press 2011); and *Foucault, Governmentality and Critique* (Paradigm 2011).

Ursula Naue is Senior Lecturer in the Department of Political Science at the University of Vienna in Austria, and is a political scientist by training. Her research and teaching focus is on inclusion, exclusion and marginalisation, especially related to disability policies and politics, and the politics of dementia and age(ing). She also teaches courses in comparative politics, policy analysis, governance analysis, and Disability Studies.

Anna-Maria Tapaninen is Research Fellow in the Department of Social Sciences at the University of Eastern Finland. Her doctoral dissertation in social anthropology, based on ethnographic fieldwork, was on women's everyday life in a South Italian housing estate. The monograph (in Finnish) was entitled *Kansan kodit ja kaupungin kadut: Etnografinen tutkimus eteläitalialaisesta kaupungista.* She has also carried out archival research on institutionalised child abandonment in nineteenth-century Europe, Naples in particular. She is currently studying the use of biotechnologies in the context of immigration in Finland, with a focus on DNA testing and medical age assessment.

Martin G. Weiss is Assistant Professor in the Department of Philosophy at the University of Klagenfurt, Austria. He collaborates with the Interdisciplinary Research Platform Life-Science-Governance of the University of Vienna. He is particularly interested in bioethical and biopolitical implications of biotechnologies. Recent publications include *Bios und Zoë. Die menschliche Natur im Zeitalter ihrer technischen Reproduzierbarkeit* (Suhrkamp 2009) and *Ethics, Society, Politics* (co-edited with Hajo Greif; De Gruyter 2013).

Acknowledgements

The project 'DNA and immigration: Social, political and ethical implications of DNA analysis for family reunification' (IMMIGENE) was funded by the Academy of Finland (grant number 135266), the Austrian Research Promotion Agency (grant number 823817) and the German Federal Ministry of Education and Research (grant number 01GP0903).

We would like to thank, first and foremost, our interviewees and informants who shared their views and experiences regarding DNA analysis for family reunification with us. Without these valuable insights, this project would not have been possible. We would also like to thank all research assistants involved in the project: Sarah Dionisius, Elina Helosvuori, Juliko Lefelmann, Caroline Mitschke and Shirin Moghaddari. Our project benefitted greatly from valuable discussions with and helpful comments received from our colleagues, especially Amade M'Charek, Barbara Prainsack, Katharina Schramm, Susanne Schultz, David Skinner and the participants in the final IMMIGENE workshop in Frankfurt on 22–23 November 2012. We would also like to express our gratitude to Carolin Mezes who helped with preparing the manuscript and to Gerard Holden who copy-edited the text. Last but certainly not least, we thank Gemma Hayman and Neil Jordan of Ashgate for their continuous support and help during the editorial process and in the production of the book.

Chapter 1
Constellations, Complexities and Challenges of Researching DNA Analysis for Family Reunification: An Introduction

Torsten Heinemann, Ilpo Helén, Thomas Lemke, Ursula Naue and Martin G. Weiss

Almost 30 years ago, in 1985, the British geneticist Alec Jeffreys made a discovery that soon caused a lot of excitement and became famous as the idea of a 'genetic fingerprint'. This technology was rapidly used in the context of crime scene investigations, criminal identification and law enforcement. It is less well known, though, that the first site where this new technology was applied was not a criminal case but unresolved family issues. Soon after Jeffreys and his colleagues published two articles on their discovery in *Nature* (Jeffreys, Wilson and Thein 1985a, b) and the popular press reported their remarkable findings, a team of lawyers got in touch with him. They were advising

> a Ghanaian boy born in the United Kingdom who emigrated to Ghana to join his father and subsequently returned alone to the United Kingdom to be reunited with his mother, brother and two sisters. However, there was evidence to suggest that a substitution might have occurred, either for an unrelated boy, or for a son of a sister of the mother. As a result the returning boy was not granted residence in the United Kingdom. (Jeffreys, Brookfield and Semeonoff 1985, 818)

Jeffreys took blood samples from the boy, the alleged mother and the alleged three siblings. The DNA analysis of the samples showed that the boy was indeed the biological son of the mother and a full sibling of her other son and two daughters. On the basis of the DNA evidence, the Home Office acknowledged the existing family relation and the boy was allowed to stay in the UK. This case became the starting point for a wide application of DNA analysis in immigration cases:

> Over the next decade, DNA fingerprinting was used to test more than 18,000 immigrants who had been refused entry into the UK. Of these, more than 95% produced results that showed they were blood relatives of UK citizens and were

> therefore entitled to British citizenship – thanks to DNA fingerprinting.[1] (McKie 2009, 16)

Since the 1990s, many countries around the world have followed the United Kingdom's lead and started to use DNA testing to verify genetic relatedness in family reunification cases.

In general, family reunification refers to the right of foreign family members living abroad to join relatives who hold long-term residence permits or are citizens of a given country. This right has been an integral part of many countries' immigration policies, and is meant to protect the family in accordance with the Universal Declaration of Human Rights (UDHR) and the European Convention on Human Rights (ECHR).

The current trend among host countries seems to favour more restrictive family reunification policies. Many countries around the world are imposing stricter requirements on those applying to reunite with their family members. They often ask applicants to provide official documentation to prove their identities. Providing such information is often difficult, especially in countries that do not use official documents to establish identity or where those documents have been lost or destroyed due to politically unstable situations. Even if applicants possess the required documents, the information is sometimes rejected by immigration authorities as they question the authenticity of the documents. In this context, many countries resort to DNA kinship testing to resolve cases in which they consider the information presented on family relations to be incomplete or unsatisfactory (Taitz, Weekers and Mosca 2002b; Heinemann, Naue and Tapaninen 2013).

Although exact usage rates and statistics are unavailable in most host countries, there is evidence that DNA testing in the context of immigration policies is on the rise. Today, at least 21 countries around the world including 17 European countries have incorporated parental testing into decision making on family reunification in immigration cases: Australia, Austria, Belgium, Canada, Denmark, Estonia, Finland, France, Germany, Hungary, Italy, Lithuania, Luxembourg, Malta, the Netherlands, New Zealand, Norway, Switzerland, Sweden, the UK and the USA (European Migration Network 2008, 2009b; Heinemann and Lemke 2013; La Spina 2012).

DNA analysis certainly offers some technical advantages over traditional methods of identification such as blood testing (Davis 1994), as it is much more precise, less prone to errors, fairly cheap, and provides a basis for establishing genetic relatedness not only between parents and their children but also between siblings or cousins or between children and their grandparents, aunts and uncles or nieces and nephews. In other words, it allows the examination of complex and extended family relations and achieves very high probability rates (Blouin 2003; Jones and Ardren 2003; Jones et al. 2010). DNA testing might also be considered an effective instrument to prevent child-trafficking and to limit fraudulent family

1 For the history of DNA profiling, see Cole 2001.

reunification (European Commission 2011; Müller 2012). While parental testing is a very precise way of determining biological relatedness, the technological procedure also has some limits: for example, children may have natural mutations that lead to an exclusion of maternity or paternity even though a biological relation between the parents and the child exists (see Junge et al. 2006). Furthermore, human errors can occur both in taking samples in the countries of origin and when analysing them (see Dawid et al. 2001; Karlsson et al. 2007; Mansuet-Lupo et al. 2009).

While these issues are primarily technical and can be solved or dealt with in practice (Heinemann, Naue and Tapaninen 2013; Heinemann and Lemke 2014), DNA testing for family reunification in immigration cases also raises more serious concerns that have not so far been adequately addressed. It has far-reaching consequences not only for immigration policy, but also for the idea of citizenship, privacy issues, and the concept of family. In legal terms, the question arises of how the applicant's right to privacy and family is respected (Frenz 2008). A related issue concerns data protection: how long is DNA information stored, and what agencies have access to it? There are also serious social concerns that go along with DNA testing in the context of family reunification. The focus on genetic links may devalue alternative forms of family that are seen as secondary or even excluded. Only biological relatives would be eligible for family reunification. A further aspect to be considered is the fact that some families will not be able to provide DNA test results, since they either lack the financial resources to pay for the test or religious convictions will make it impossible for them to consent to the procedure (Taitz, Weekers and Mosca 2002a, 2002b).

DNA testing for family reunification also poses ethical questions, as this practice potentially conflicts with the concept of informed consent and the right not to know, two fundamental principles of medical ethics. In addition, the use of DNA tests in immigration policy may violate the right of the individual to decide what information about himself/herself may be made available to others and under what circumstances. This implies that DNA tests must be absolutely voluntary, which might be doubted in the context of DNA testing for immigration purposes since a person choosing not to submit to the test would probably lose the opportunity to be reunited with his/her family.

Many journalists around the world have reported on individual cases in which the use of DNA analysis for family reunification has raised serious legal, ethical and social concerns (see for example Funk 2007, 2009; Gaserow 2007; Swarns 2007; Weiss 2011). Beyond this media coverage, research and scientific publications in this field are rare. So far, the topic has only stirred interest among legal scholars and in the human rights literature (Taitz, Weekers and Mosca 2002a, 2002b; Frenz 2008; Murdock 2008; Holland 2011). Social science scholars have mostly focused on the use of parental testing and its implications for family relations, parenthood, and gender aspects (for example Turney 2005, 2006, 2011; Freeman and Richards 2006; Richards 2010; Fonseca 2011) within one country. While there is a body of research on the social, political and ethical implications of genetics for health and medicine, agriculture and nutrition, and insurance and employment, the impact of forensic genetics on immigration policies and family reunification has only recently become an object of interest for social scientists (Dijstelbloem and Meijer 2011; Heinemann, Lemke and Prainsack 2012; Heinemann and Lemke 2013, 2014; Heinemann, Naue and Tapaninen 2013; Tutton, Hauskeller and Sturdy 2014). So far hardly any research has empirically investigated different national regimes of using DNA analysis for family reunification in a comparative perspective, or systematically addressed the normative issues at stake.

This volume is the first contribution in the field to provide a systematic analysis of the ethical, political and social challenges resulting from DNA testing for immigration purposes. The comparison between Austria, Finland, and Germany helps us to better understand which factors determine the trajectories and the outcomes of family reunification processes. It generates knowledge on how DNA information is used and mobilised in decision making on family reunification, and it also improves our understanding of the historical and institutional dynamics of immigration policies.

The volume has emerged from the international project 'DNA and immigration: Social, political and ethical implications of DNA analysis for family reunification' (IMMIGENE). Most of the empirical research took place in 2011 and 2012. The project consortium consisted of researchers from Austria, Finland and Germany. Kevin Hall and Ursula Naue were responsible for the Austrian project, Anna-Maria Tapaninen and Ilpo Helén did the research in Finland, Torsten Heinemann and Thomas Lemke were in charge of the German project, and Martin G. Weiss and Jakob Guggenheimer explored the ethical implications of DNA testing in family reunification.

In this introductory chapter, we first outline the European framework of family reunification and the reasons for focussing on Austria, Finland and Germany. After that, we spell out the theoretical debates and the methodological challenges involved in researching the use of DNA analysis in three different countries. Finally, we provide an overview of the structure of the volume.

Family Reunification in the European Union: An Overview

Family-related immigration is currently the most important form of legal immigration to the European Union (European Migration Network 2008; Honohan 2009; Eurostat 2011; Ruffer 2011; Müller 2012). The right to family reunification is related to the protection of the family as laid down in article 16 of the UDHR and article 8 of the ECHR (Feller et al. 2003, Jastram and Newland 2003; Heinemann, Naue and Tapaninen 2013). Article 8 of the ECHR protects individuals from unlawful interference in 'one's private and family life, home and correspondence'. The Convention also establishes the European Court of Human Rights with the authority to give binding judgements to the state parties. In its judgements, the court has 'recognized that article 8 places State Parties under a *positive* obligation to admit a family member in a situation where the family unit cannot be reasonably expected to relocate in the country of origin' (Cholewinski 2002, 275, emphasis by authors). In the European Union, the right to family is laid down in the Charter of Fundamental Rights of the European Union, which stresses the right to have a family in two sections: in article 7 on respect for private and family life, which is based on article 8 ECHR, and in article 9 on the right to marry and to found a family. However, beyond the general scope of these articles the charter 'is silent on the question of family reunification and rather muted on the question of protecting the family' (Cholewinski 2002, 276; see also John 2003).

The Treaty on the Functioning of the European Union (TFEU) states in article 79 that 'the Union shall develop a common immigration policy'. Article 79(2) a further specifies that the European Parliament and the Council shall adopt necessary measures in the areas of 'the conditions of entry and residence, and standards on the issue by Member States of long-term visas and residence permits, *including those for the purpose of family reunification*' (emphasis by authors). The details, consequences and administrative practices are not specified further, but the treaty defines a minimum standard that has to be guaranteed by all EU Member States. In the end, it is up to each Member State to implement procedures that are in line with the treaty (Schibel 2004).

Council Directive 2003/86/EC, ratified in 2003 by all EU Member States except Denmark, Ireland and the UK, is the most important European legal document in regard to family reunification. It contains legal information on the rights of the applicant and the family, and provides a framework for administrative practice. However, this framework leaves considerable room for interpretation regarding the definition of the family, the right of minors to apply for family reunification, the age limits for minors and spouses, the status of marriage as a prerequisite for the application, and certainly the use of DNA analysis to verify family relationships (Schibel 2004). Article 5(2) only states that 'the application shall be accompanied by documentary evidence of the family relationship' and 'if appropriate, in order to obtain that a family relationship exists, Member States may carry out interviews … and conduct other investigations that are found to be necessary'. One of those 'other investigations' or forms of documentary evidence may be the result of a DNA kinship test, but this is not specified.

While this scope for interpretation may be seen as a problem in establishing a common immigration and family reunification policy, it is important to emphasise that Council Directive 2003/86/EC does not establish more requirements for family reunification than the regulations and laws governing family reunification in the Member States (International Federation for European Law 2004; Schibel 2004). Thus, Council Directive 2003/86/EC does not actually guarantee new rights for those applying for family reunification but instead tries to align them among the Member States, and does this at a (sub)standard level. Most EU Member States seem to prefer a rather strict definition of the family and allow family reunification only to married couples and their children under the age of 18, or sometimes even 16 (European Migration Network 2008, 2009a, 2010). Notwithstanding justified criticism, it has to be noted that '[t]he Directive has not only extended basic rights and legal securities to new immigration countries, but also secured them from future policy restrictions in all countries' (Huddleston 2011, 2).

In 2011, the European Commission initiated a public debate on the Directive through a Green Paper in which all stakeholders, that is to say the Member States, NGOs and any individuals who might want to comment were invited to respond to a number of questions and discuss issues related to the Directive. These questions addressed, firstly, the shortcomings in the harmonisation and implementation of the Directive and, secondly, the restrictive turn in migration policy in general that had seen calls for amendments aimed at establishing additional conditions for family reunification (European Commission 2011; Huddleston 2011). In fact, after the transposition of the Council Directive 2003/86/EC many Member States introduced stricter regulations for family reunification. The restrictive turn was mirrored in the first European Migration Network Focus Study, which was meant to inform the Green Paper in relation to marriages of convenience and false declarations of parenthood. The title of the report was *Misuse of the Right to Family Reunification* (European Migration Network 2012a). Both in the Focus Study and the Green Paper, the use of DNA analysis in immigration management was explicitly linked to the possible fraudulent use of the right to family reunification.

Austria, Finland and Germany represent three distinct ways of adopting EU regulations for family reunification into national law. In 2000, and as one of the first European countries to do this, Finland adopted DNA testing for immigration purposes and developed specific guidelines and legal provisions to regulate this practice. The common regulations for family reunification as laid down in Council Directive 2003/86/EC had to be transposed into national law by 3 October 2005, which resulted in the introduction of several new laws regulating immigration and family reunification in all Member States. Austria and Germany introduced the new immigration laws in 2005, while Finland anticipated the provisions of the Council Directive 2003/86/EC in the new Aliens Act (Ulkomaalaislaki, UlkL 301/2004) in 2004 and subsequently made further amendments in 2006 (SDK 380/2006) to fully transpose the Directive.

Not only do Austria, Finland, and Germany differ in when and how they incorporated the Council Directive into national law, they also represent very different regimes for using DNA analysis in decision making. German immigration authorities perceive the results of a DNA test as a highly credible and crucial piece of information that trumps all other evidence and will ultimately allow for family reunification if it shows that family members are indeed biologically related. Austria uses the test primarily to verify or falsify statements made by the applicants in interviews. If the test confirms the statements of the applicants in the interview, immigration authorities assume that they can trust the applicant in general. It thus functions as a sort of lie detector, and tests the credibility of the applicant. Finland has a different regime. On the one hand, DNA testing is considered to secure the rights of immigrants in the family reunification process in Finland, and therefore the use of DNA analysis is meticulously regulated by the authorities. On the other hand, Finland enforces a rather strict policy for applicants in an effort to combat fraud and misuse of the family reunification procedure. These substantial differences make the comparison between the three countries particularly interesting, and offer new insights into the divergent and multiple ways in which DNA tests are applied in immigration management.

Theoretical Contexts

Our inquiry is inspired by two approaches from social research which focus on interconnections between scientific research and practices of governing people in Western societies today: first, ethnographic studies and sociological investigations on the use of genetic knowledge in non-medical contexts such as forensic investigation and insurance (for example Hindmarsh and Prainsack 2010; Lynch, Cole, McNally and Jordan 2008; M'charek 2005); and second, Foucauldian analytics of government (for example Bröckling, Krasmann and Lemke 2011; Dean 2010; Helén 2005). By combining these perspectives, our study of the enactments and the practices of DNA testing in decision making on immigration and family reunification can show essential features and important transformations of policy-making and administrative practice. The Foucauldian perspective in particular directs our analysis to focus on *problematisation in practice*. We concentrate on the ways in which certain issues are seen as problems and worked out, and on contexts of reasoning and practice in which the problems emerge, are defined and attempts are made to solve them (Helén 2005).

With this theoretical and analytical focus, the findings presented contribute to an important discussion in social studies of biotechnologies and the new genetics. The volume seeks to contribute to and complement the debate on 'biological citizenship' (Petryna 2002) or 'genetic citizenship' (Kerr 2003). The growing literature on this topic mostly refers to the importance of patients' associations, disease advocacy organisations and self-help groups that are giving rise to new forms of subjectivation and collective action by challenging existing borderlines between laypeople and scientific experts and between active researchers and passive beneficiaries of technological progress. Frequently, authors using the notion of biological or genetic citizenship argue that these groups are questioning access to knowledge and claims to expertise, forging new alliances with biomedical researchers, and lobbying to influence political decision making and to receive funding for medical research (Heath, Rapp and Taussig 2004; Rose and Novas 2005; Rose 2007; Fitzgerald 2008; Schaffer et al. 2008).

However, there are other important aspects of biological/genetic citizenship that need to be explored. One central question that has so far been neglected addresses the link between biology and citizenship, asking what biological prerequisites an individual must fulfil in order to become a citizen (Rose and Novas 2005). The use of DNA information in immigration decisions and the proof of genetic ties in family reunification promises to open up a new dimension in the discussion of genetic citizenship, as up until now the debate has concentrated on the medical sphere. By investigating a new empirical field, we will argue that a hitherto neglected dimension of the link between biology and citizenship becomes especially pertinent in the context of DNA testing for family reunification in immigration processes.

Hence, the volume seeks to complement and redirect the debate on biological citizenship by expanding the analytical focus, following new empirical questions and inquiring into different areas of research.

Comparative Analysis as a Challenge: Reflections on Methodology

Comparative analysis of the use of DNA testing for family reunification poses many methodological and practical challenges. Parental testing in general, and in family reunification cases in particular, is a sensitive and controversial topic. Therefore, data collection and analysis in this context need special planning and attention. Furthermore, comparative analysis of this practice is particularly challenging because it takes place in a complex set of international and national legislations, and administrative practices differ considerably between the three countries under investigation as well as between the EU Member States. Thus a simple comparison of one country with another is not a suitable approach.

The project findings presented in this volume are based on a qualitative study involving 'multisite research', consisting of document analysis and interviews within a Grounded Theory framework (Clarke 2005, 145–146). Methodologically, the research project was guided by the interpretative approach to the analysis of public policy and governance (Yanow 1996, 2000; Fischer 2003), and the research strategy and design was characterised by a qualitative social-scientific case study approach (see Yin 2014).

The empirical data mainly consist of two bodies of material. In a first step, we collected and analysed key documents and reviewed literature on DNA testing for family reunification. Relevant documents are statutes, draft laws, guidelines, directives, parliamentary debates, official meeting minutes of hearings and public debates, government and expert committee reports, policy statements and protocols, NGO materials and media reports. Additionally, we looked at court hearings and decisions in legal cases. The reviewed literature consists of articles and books dealing with the social, political, legal and ethical implications of DNA testing in the context of immigration management, as well as medical and scientific literature focusing on the technical challenges of parental testing.

Building on the findings of the document analysis and literature review, in the second step we conducted semi-structured, in-depth qualitative interviews with actors in the field of immigration management relevant for family reunification process – scientific experts, administrative personnel and policymakers, civil society actors and immigrants (Bogner, Littig and Menz 2009). We interviewed representatives of international governmental organisations, international and national NGOs and immigration authorities, lawyers specialising in immigration law, geneticists, and applicants for family reunification. In total we conducted 76 interviews in the three countries during 2011 and 2012.

In order to align our methodological approach and to improve the comparability of our data, we developed joint interview and interpretation guidelines. These guidelines were adapted to the national context in order to be able to address different issues in the three countries. On average, the interviews lasted 90 minutes. For the analysis, the interviews were transcribed, anonymised, and then interpreted using a qualitative content analysis approach (Flick 2007, 2009; Gibbs 2007; Mayring 2004). In the following chapters, direct and indirect quotations from the interviews will be referenced with the respective function of the interviewee and a consecutive number of the interview, for example 'German immigration officer 1' or 'Geneticist 2'.

In many ways, finding informants, conducting interviews and gathering material has been akin to fieldwork. In this sense, our research follows the methodological approaches of Grounded Theory and ethnographic research (for example Birks and Mills 2011; Clarke 2005; Hammersley and Atkinson 2007). Embedded in and directed by our data, our analysis attempts to capture the use of DNA profiling for family reunification as it is understood, conceptualised and reasoned over by those involved in actual practices within the immigration management system in Austria, Finland and Germany. We consider our study a process in which a strictly structured approach would not yield valuable results. Instead, the interviews and encounters are illuminating takes of evolving narratives within the field. In what follows we do not specify the data quantitatively or analyse any distinct views attributable to the institutions in question. Nor do we decipher the attitudes or opinions of single informants or stakeholders. Rather, we understand our data in the context of dialogically unfolding discussions in a dynamic and contested field in which the story of DNA testing is narrated and the rationales and justifications for the procedure are formulated by the authorities and experts involved.

Structure of the Volume

In the following chapters, we analyse the current legal framework in Austria, Finland and Germany and the social, political and ethical implications of DNA testing for immigration purposes in those countries. The case studies provide a general understanding of how DNA testing affects the national immigration policies and identify emerging cross-national patterns of the use of DNA evidence for immigration management. These empirical findings are complemented by a substantive normative investigation assessing the ethical implications of DNA testing for family reunification and how they should be addressed in immigration policies.

Chapter 2 investigates the introduction and growing application of parental testing for family reunification in Germany. In their analysis, Torsten Heinemann and Thomas Lemke pay special attention to the implicit model of the family and kinship that governs the German immigration regime. They show that German immigration authorities rely on DNA evidence as the ultimate proof of an existing family relation. The authors argue that DNA analysis for family reunification strengthens a biological family model which contrasts with the more pluralistic and social concepts of family in Germany as well as in many societies in Europe and North America, and results in a double standard for family recognition for Germans and for immigrants.

In Chapter 3, Anna-Maria Tapaninen and Ilpo Helén situate DNA testing in the context of the Finnish immigration regime in the late 1990s and early 2000s. Finland was one of the pioneers in the adoption of this technology, and the implementation of DNA testing in the Finnish immigration administration is analysed on the basis of primary documents and expert interviews. Tapaninen and Helén also discuss policy dynamics within the current, increasingly restrictive immigration regime and point out two policy rationales which frame the use of DNA analysis for family reunification: securing human rights and combating fraud. Finally, they focus on the decision-making procedure in which evidence of genetic kinship ties on the one hand, and evidence of genuine family life on the other, are weighed together in ambivalent ways.

Chapter 4 investigates the situation in Austria. Kevin Hall and Ursula Naue show that the Austrian authorities triangulate different types of evidence when dealing with applications for family reunification. As the main targets of DNA testing, refugees and their families are required to provide documents and attend interviews at various stages throughout immigration procedure. These statements become the focus of the authorities' scrutiny. The authors argue that in this context, a DNA test acts in a double sense as a 'lie detector': to verify the family status, and also to verify information provided in documents and/or interviews with authorities. This chapter describes the procedure, and explores some of the implications of DNA testing in family reunification cases in Austria within the given legal framework and political practices.

The use of DNA analysis for family reunification generally raises a number of philosophical questions and ethical concerns. This topic is a battleground of conflicting values and interests, which are discussed by Martin G. Weiss in Chapter 5. The ethical problems addressed here are the structural (im)possibility of giving informed consent to the DNA test, the right to know and not to know, the special sensitivity needed in the treatment of underage children within the reunification process, and the question of whether the use of DNA testing to verify family ties may not be less intrusive, as far as privacy is concerned, than the traditional alternative of distressing interviews about applicants' private life.

Chapter 6 sums up the findings of the comparative analysis and the philosophical investigation. It addresses the two theoretical debates mentioned in this introduction, and relates the analysis of DNA testing to the geneticisation thesis and to the debate on biological citizenship. It also engages with the interpretation of DNA as a lie detector and truth machine. Finally, we offer some recommendations and reflections for future practice drawing on an investigation of the current regulatory framework and governance regime, an understanding of the situation in other European and non-European countries, and the analysis of the societal, political and ethical issues.

Chapter 2
Germany: The Geneticisation of the Family[1]

Torsten Heinemann and Thomas Lemke

Introduction

Family reunification is the statistically most significant form of legal immigration in Germany, and in the past 10 years it has outnumbered positive decisions on asylum and any other ground for immigration (Kreienbrink and Rühl 2007; European Migration Network 2008, 2011; Müller 2012).[2] It comes as no surprise, therefore, that the topic has received a considerable amount of attention from policy makers and in public debate. The role of DNA analysis in family reunification, however, was a non-issue for political and media actors until 2007. This changed when, in that year, the French president Nicolas Sarkozy proposed a much more restrictive immigration law including the use of DNA tests to verify family relatedness. German newspapers were very critical in their comments on this proposal.[3] In the course of the public debate, though, they soon discovered that immigration authorities in Germany had been using parental testing for a long time to determine biological relatedness in family reunification cases. As there was until very recently no legal foundation for employing DNA evidence in immigration cases, this practice remained unnoticed by the public for a long time.

This chapter will outline the German regime for the use of DNA analysis in family reunification cases. It starts with a brief overview of the right to family reunification in German legislation and the role of DNA tests in the administrative process. The next section presents evidence that DNA testing is increasingly being employed in administrative decision making on family reunification in Germany. We then highlight important societal, political and legal implications of the use of parental testing in family reunification. Firstly, there is a tension between the formal voluntariness of DNA testing in family reunification and

1 A version of this chapter has been published in *Sociology* 47(4), August 2013, by SAGE Publications Ltd, all rights reserved. © Torsten Heinemann and Thomas Lemke, 2013. For the purpose of the present volume, the text has been expanded, substantially revised and updated.

2 In 2012, 31,126 positive decisions on applications for or the renewal of asylum were taken, while 54,816 residence permits were granted for the purpose of family reunification (Federal Office for Migration and Refugees 2013, 45, 89).

3 The headlines in the press read, for example, 'France discovers migration gene' (Hahn 2007) or 'Biology instead of legal rights' (Schmidt 2007).

the actual necessity to provide DNA evidence in the administrative procedure. Certain groups of applicants, in particular those originating from countries in sub-Saharan Africa or South East Asia, de facto have to undergo parental testing as their applications will be rejected otherwise. Secondly, in the context of family reunification, important rights and legal guarantees that protect individuals from the abuse of their genetic information in other social and medical domains are inoperative. It is quite remarkable that applicants for family reunification have no right to decide what happens to their DNA profiles once the test has been carried out. This legal inconsistency might also promote a general suspicion towards applicants for family reunification. Thirdly, we present evidence that the use of parental testing in decision making on immigration endorses a genetic concept of family and kinship. It establishes a 'double standard of family recognition' (Murdock 2008), as the requirement of a biological link between migrant family members is diametrically opposed to family recognition policy for German citizens, which emphasises social rather than genetic ties. In the final part of the chapter we argue that the German legal framing and the administrative practice of DNA testing for family reunification result in a geneticisation of the family. Particularly in contrast to Finland (see Chapter 3), the focus on genetic links tends to devalue social forms of family that are seen as secondary. The DNA test serves as a means to verify family relatedness, and proof of biological ties provides a solid basis for administrative reasoning and almost automatically leads to a positive decision on the application.

The argument of this chapter is based on an extensive literature and document analysis, as well as interviews with relevant actors involved in decision making on family reunification and DNA analysis (see also Chapter 1 for details of the methodology). We reviewed and analysed documents that deal with parental testing in immigration contexts such as laws, guidelines, directives, minutes of parliamentary debates, government and expert committee reports, NGO materials and media reports. Additionally, we conducted semi-structured in-depth interviews with actors in the field (Bogner, Littig and Menz 2009): five experts working for German and international NGOs, a senior officer working for the UN High Commissioner for Refugees (UNHCR), five officers in German Aliens Departments and immigration authorities, four lawyers specialising in family and immigration law, and five geneticists or forensic doctors in genetic and forensic laboratories. We also interviewed seven applicants for family reunification who underwent a DNA test during the application procedure.

Family Reunification in Germany: Legal Framework, Historical Determinants and Administrative Practice

In Germany, the most important piece of legislation with regard to family reunification is the 2005 Residence Act (Aufenthaltsgesetz, AufenthG) as part of the Immigration Act (Zuwanderungsgesetz, ZuwandG). The Residence Act, which replaced the Aliens Act, incorporated most of the regulations of Council Directive 2003/86/EC and adapted them to national law. Part 6 (articles 27–36) is specifically concerned with family reunification. The Residence Act explicitly states that the right to family reunification is meant to protect the family in accordance with the German Basic Law (Grundgesetz, GG).

In accordance with the Residence Act every spouse, either a German or a foreigner who is in possession of a temporary or unrestricted residence permit, can be the sponsor of an application for family reunification. The sponsor and his/her partner need to be married. In principle, same sex partners may apply for family reunification as well and they have the same rights as married couples. However, the same sex partnership or marriage has to be officially recognised as such in the country of origin. For many applicants it is almost impossible to fulfil this prerequisite, as same sex partnerships are not legally acknowledged in many home countries. Even worse, the fact that someone is gay, lesbian, bi- or transsexual may well have been the main reason for leaving the country of origin as this person may have faced persecution because of their sexual orientation. It might also be a valid foundation for a successful application for asylum in the EU, only for the person concerned to find out that they cannot be reunited with their partner.

In addition to the prerequisite of an existing marriage or recognised same-sex partnership, the sponsor is also expected to provide evidence of an adequate income and enough living space for the prospective united family. Accepted asylum seekers and refugees do not need to provide evidence of a living wage and living space if they apply for family reunification within three months after they have been officially recognised by the Federal Office for Migration and Refugees. Children holding a temporary or unrestricted residence permit may also serve as sponsors and apply to be reunited with their parents. However, the provisions in the Residence Act generally assume that the person serving as a sponsor is an adult.

The right to family reunification is legally recognised only for spouses over 18 years old and their underage children (European Migration Network 2010a, 2010b; Huber 2010). Thus, the Residence Act establishes a nuclear family model. Only under so-called extraordinary hardship can the right of residence also be granted to more distant relatives such as grandparents, children who have reached adulthood and siblings of any age (Walter 2009). However, in 2012 only 243 residence permits for family reunification were issued in these cases of exceptional circumstances, which was 0.4 per cent of the total number of residence permits granted for family reunification (Federal Office for Migration and Refugees 2013, 89).

These fairly strict regulations have to be interpreted in the German historical context by taking into account the immigration wave in the 1960s and 1970s, which occurred as a result of the German 'economic miracle' (Wirtschaftswunder). During that time, Germany actively recruited foreign workers as so-called *Gastarbeiter* from the Mediterranean countries. Between 1961 and 1971 the number of foreigners living in the country went up from 700,000 to almost 3.5 million, most of them being Gastarbeiter. This was the most significant immigration wave Germany had seen in its history (Bade and Anderson 1994,

90). Following the economic slowdown in the 1970s, the recruitment of foreign labour came to a halt. After the German government introduced the Anwerbestopp in 1973, family reunification was the most important form of immigration because a significant number of Gastarbeiter wanted to stay and be reunited with their families in Germany. In the 1980s there was a huge public and political debate about how to cope with this family-related influx of migrants. As an outcome of these discussions, Germany introduced a strict policy on family reunification in its amendment of the Aliens Act in 1990. The amendment codified the nuclear family as a model for family reunification and set narrow guidelines for the maximum age of children to be reunited with their parents. It was also around this time that DNA analyses were first used in family reunification cases in Germany (Geneticist 2).

Concerning the administrative procedure, it has to be noted that the application for family reunification is filed at a German embassy. In contrast to Austria and Finland (see the following chapters), the application can be submitted either by the non-EU foreign national or by the so-called sponsor, that is to say the person holding the residence permit in Germany. In the vast majority of cases, the applicants engage a lawyer in Germany who takes care of the rather complicated administrative procedure.[4] The embassies are in charge of the entire application process, and will also make the final decision on particular cases. The Aliens Departments in the urban districts of the applicant's place of residence in Germany may be contacted by the embassy to check whether the applicant can provide enough living space and the required living wage for the prospective united family. All relevant documents will be finally checked and verified by the embassies. Family members who apply for family reunification in Germany have to establish their family status by presenting official documents or other suitable pieces of evidence. In contrast to other countries such as Austria and Finland, the whole process is entirely based on documentary evidence. A lawyer specialising in immigration law mentioned in an interview that even without DNA analysis there are at least 15 possible ways to prove a family relationship in German immigration law: certificates of birth, marriage or death, wedding photos, family and holiday photos, statements in the asylum procedure hearings, sworn declarations by the applicants or by persons who know the applicants personally, common bank accounts, regular money transfers, phone calls, email exchanges and internet chats, or sound and full particulars about the family relations made in the course of the official interview for the asylum application (Lawyer 2).

However, immigration officers at the German embassies and in the Aliens Departments in Germany will not necessarily accept these pieces of evidence, either because they consider them insufficient to prove an existing family relation or because they suspect fraud. There has been press coverage of two cases where more than ten documents were provided to but not accepted by the German

4 In all cases we know of, there was only one where no lawyer was involved, and this person declared after she went through the administrative procedure: 'If I had to do it again, I would immediately engage a lawyer to facilitate the process' (Applicant 7).

Federal Foreign Office (Gaserow 2007; Funk 2008), which is in charge of the German embassies and thus responsible for issuing visas for applicants.[5] Even in cases where legal documents are presented, it is a common administrative practice to additionally ask the applicants for a DNA kinship report (Kreienbrink and Rühl 2007; Federal Government of Germany 2008; Frenz 2008; European Migration Network 2010b). Moreover, the German Federal Foreign Office has published a list of over 40 countries whose documents are not acknowledged by German embassies at all because they assume that their system of identity registration lacks systematic and sound procedures (Federal Government of Germany 2008).[6] Immigrants from these countries will find it extremely difficult to prove a family relationship by official documents. To obtain permission to reunite with family members, they generally have to resort to DNA testing. German citizens may be asked to provide DNA evidence for their biological relation as well, if they apply for family reunification with a foreign spouse and children from one of these blacklisted countries. But the request to perform a DNA test may go beyond cases that concern citizens of blacklisted countries (see also Heinemann, Naue and Tapaninen 2013). Some of our interviewees reported that German citizens were asked to submit the results of a DNA test as a proof of a biological family relation after they applied for family reunification with family members from other non-EU countries such as Turkey (NGO officers 1, 2).

Parental testing for family reunification in Germany concerns four different groups of applicants. Firstly, there are Germans who want to reunite with their non-EU spouses and children not in possession of a valid residence permit. Almost fifty per cent of all residence permits for family reunification were issued to individuals who reunited with a German partner (Federal Office for Migration and Refugees 2013, 89). German immigration officers therefore refer to cases in this group as 'standard cases' (Immigration officers 1–3). DNA tests are only rarely used in these cases, but as we will show below this does happen under certain circumstances. The second group consists of non-EU foreign nationals holding a residence permit, who are usually recognised refugees who want to reunite with their spouses and children (Immigration officers 1–4; Lawyers 1, 2, 4; NGO officers 1–4, UNHCR senior officer). Thirdly, there are bi-national parents who are neither married nor partners but who want to stay together in Germany to jointly raise their common children. In these cases, the right of the non-EU foreign national parent to obtain a residence permit is dependent on the relation to the child, and in such cases the immigration authorities tend to question the parenthood (Immigration officers 1, 2; NGO officers 2, 4; Lawyer 3). Finally,

5 In the end, the German Federal Foreign Office and the families agreed in court hearings to settle the case: the families were allowed to enter Germany but agreed to undergo a DNA test to prove their family relations.

6 The latest information on the countries included in the list is available on the website of the German Federal Foreign Office: http://www.konsularinfo.diplo.de [accessed 10 March 2014].

there are the ethnic German resettlers and descendants of German Jews who were expelled during the Nazi era who want to return to Germany under the Federal Expellee Law. The most common example from this group is a male person who receives a residence permit for Germany in accordance with the Federal Expellee Law and has a child born out of wedlock. In these cases the father's name in often not mentioned in the child's certificate of birth, and hence the DNA test is a way to prove the family relation (NGO officer 2).

Exceptional Piece of Evidence or Standard Tool?

DNA kinship testing is explicitly mentioned in the general administrative regulations for the Residence Act (no. 27.0.5 AVwV AufenthG). The Federal Government of Germany and the Federal Ministry of the Interior stress that DNA tests are not to be seen as a constraint but as an opportunity for the applicants to prove the validity of their application (Federal Government of Germany 2008). Furthermore, they emphasise the voluntary character of the DNA tests and argue that it is up to the applicants whether they wish to take up this option (Federal Government of Germany 2008; see also 27.0.5 AVwV AufenthG). Finally, the authorities point out that DNA analyses are only used as a last resort to establish family links required if all other possible ways of verifying family relatedness have been exhausted. As an immigration officer we interviewed put it: 'Normally, we just check the documents and that's it. We really only ask for the DNA test in very rare cases, where no legal documents can be provided' (Immigration officer 2).[7] Similarly, the German Federal Foreign Office stated in response to an inquiry by the authors that the 'DNA test for family reunification is not a standard but only an exceptional case and … *is only offered to the applicants if evidence relevant to the issue cannot otherwise be provided*' (Immigration officer 4, emphasis by the authors).

While in these statements DNA testing for family reunification is presented as an *ultima ratio*, other sources describe it as a *standard tool* for the verification of a family relationship in immigration cases (UNHCR 2008; NGO officer 2; NGO officer 3). The head of the Aliens Department of a major city in Germany declared in a written statement that 'while there is no obligation for applicants even from countries with an insufficient official documentation system to prove family relation by DNA evidence, parental testing is an *appropriate and frequently used tool of verification*' (Immigration officer 5, emphasis by the authors). A senior UNHCR officer mentioned in an interview that 'we observe an inflationary use of DNA analyses for family reunification for refugees from Africa and South-East Asia' (NGO officer 3). Similarly, a refugee advisor from a church information centre stated that in 2010 alone she had supervised more than 20 cases of Somali

7 All quotations from the interviews in this chapter have been translated by Torsten Heinemann.

refugees who were asked to prove their family relations by a DNA test in the course of the family reunification procedure (NGO officer 4).

The contradiction between the different statements and evaluations in our experts' interviews can partly be explained by the German legal framework for immigration. Decisions on immigration and visas have to be based on a case-by-case assessment. Therefore, authorities have no legal grounds for a comprehensive and systematic use of parental tests in family reunification cases, and even if it is a common administrative practice they cannot officially confirm it. While the statements of the experts might be inconclusive, additional evidence from different sources such as parliamentary documents, media reports, statistics gathered by genetic laboratories and reports by applicants for family reunification indicates that we are witnessing a routinisation and normalisation of DNA testing in administrative decision making on family reunification in Germany.

The Federal Foreign Office and the Aliens Departments in the urban districts make no secret of the fact that DNA tests are used to determine the legitimacy of an application for family reunification. However, they do not publish any statistics on how frequently they employ this measure and it is not clear whether such official statistical information exists at all (Federal Government of Germany 2008, 2010). At present there is no reliable official information on the number of tests that have been conducted so far in the application process for family reunification. The use of parental testing for family reunification has been covered by the German press, and articles in several newspapers published some years ago, suggested that such tests were at that time already being conducted in at least 600 cases annually (Funk 2007; Gaserow 2007). Since then, the number has certainly increased. The five laboratories we visited to interview geneticists and forensic doctors alone conducted a total of approximately 900 tests for family reunification in 2010. There is a very competitive market for DNA kinship testing in general and for immigration purposes in particular, and more than 50 laboratories in Germany offer DNA analyses for family reunification.[8] Thus, there is sufficient empirical evidence to conclude that the number of DNA analyses conducted in immigration cases is significantly higher than 900 per year.

Moreover, it is important to note that the figure provided here is not identical with the number of persons involved in the application process and the visas issued on the grounds of a DNA test result. A DNA test for family reunification normally includes at least three persons: the alleged father, the alleged mother, and the child. If the test is positive and confirms the biological relatedness, two visas

8 The professional organisation for experts in parental testing (Kommission zur Feststellung der Qualifikation von Abstammungsgutachtern) lists 30 laboratories that offer DNA analyses for family reunification (see http://www.kfqa.de/verzeichnis/ [accessed 10 March 2014]). There are many more laboratories that are accredited in accordance with the ISO/IEC 17025 standard and are thus allowed to conduct parental tests which will be accepted by the authorities. The authors alone know of eight additional laboratories that are not included in the list but are officially accredited to perform DNA kinship tests.

for family reunification are granted as in most cases only one person is already living in Germany. Given that there are many cases with two, three, or even more children, the number of visas granted on the basis of a DNA test is even higher. In our interviews with the applicants for family reunification the average family consisted of five persons, which results in four visas issued on the basis of a DNA test. Therefore, it is reasonable to assume that at least nine per cent of the total of 40,210 visas for family reunification in 2010 were granted only after DNA evidence was provided (Federal Office for Migration and Refugees 2011, 91).

Voluntary Option Vs De Facto Obligation

If the immigration authorities do not recognise the documents presented by the applicants, they may 'offer' them the option of taking a DNA test to prove a biological link to family members. As the burden of proof is always on the applicants (AufenthG; AVwV AufenthG), it is up to them to find a suitable laboratory and organise the entire testing procedure. Some immigration authorities suggest a certain laboratory, but generally German authorities have to accept all laboratories that are certified in accordance with the ISO/DIN 17025 standard. The costs, approximately two-hundred euros per person tested, have to be borne by the applicant as well.[9] Once the applicant has authorised a DNA laboratory, samples will be taken from all family members. This is done in the laboratory for individuals who are already in Germany, and in the German embassies in the countries of origin by a certified person, in general a physician or a geneticist. Immigration officers in the German embassies or Aliens Departments will not normally be involved in the testing procedure at any point. The sampling includes the identification of the persons by photographs and fingerprints (for babies and small children footprints are used).

The laboratories use short tandem repeat analysis (STR analysis) for parental testing. The sample will almost always be a saliva sample. In Germany, 12 STR loci have been decided upon as the basis on which an individual genetic profile can be generated, but most laboratories use at least 16 STR loci anyway. Some laboratories even employ 21 STR markers for the standard procedure and many more if there are unexpected results that need to be checked (Geneticist 3). As the laboratories use so many markers, some geneticists involved in the procedure

---

9 The costs of the DNA tests and the fact that in Germany applicants have to pay for the test themselves have given rise to a heated debate. It is certainly correct that the parental test entails additional costs for persons who often have no or only very limited financial resources. However, it has to be noted that the costs are comparably low if one takes the overall expenses of the application into account. Apart from the costs for legal counselling, applicants have to pay the costs of fees for new passports, travel expenditure and translations. All in all the expenses might – for example, in the case of a Somali refugee – easily add up to 5,000 euros or more.

declared in the interviews we conducted that they do not even need population genetics to obtain a reliable test result (at least 99.99 per cent) (Geneticists 1 and 5). However, biostatistical analysis based on population genetics is still used in the procedure of DNA testing for family reunification (Gjertson et al. 2007). Interestingly, the attribution of the reference population is sometimes made on very broad assumptions. 'If someone wants to take the test and he has a Turkish name, and he looks a bit Turkish, than we would ask if his parents were born in Turkey and we assume he is indeed Turkish' (Geneticist 5). Categories such as ethnicity or race are thereby reinforced, even though it is not always clear whether (and in what sense) the individual really belongs to the group in question. This particularly applies to applicants from sub-Saharan Africa or South East Asia who are wrongly perceived as culturally and genetically homogeneous groups (see for example Tishkoff et al. 2009). Furthermore, the use of an inappropriate allele database may result in higher error rates (Karlsson et al. 2007, 147) and thus potentially in a negative result from the DNA kinship test.

Once all relevant samples are in the laboratory it will take two to three weeks for the results to be provided to the applicant (Geneticists 1–5). As applicants often have to wait for as long as a year before their documents are checked and verified by the authorities, this is one of the biggest advantages of DNA testing for family reunification. While it often takes a very long time before applicants are reunited with their family members, with a DNA test the decision is sometimes made within less than four months from the application to the final decision. This is also why immigration lawyers often advise their clients to undergo testing: 'We, the lawyers, are quite pragmatic in this respect. It has to go fast, [and] the DNA test is helpful in this respect' (Lawyer 2).

While the test is mostly presented as an additional and above all a voluntary option, it might be doubted how voluntary the use of DNA analysis in this context can be if the application for family reunification will be rejected otherwise. The senior UNHCR officer we interviewed argued that the voluntariness of the DNA test can be compared to the 'voluntary departure' from Germany for rejected asylum seekers who will otherwise be deported anyway (UNHCR senior officer). Here, we note a remarkable parallel to forensic DNA profiling: 'Persons who refuse to give a "voluntary" sample (which is their legal right) attract police attention and become more suspicious as a result' (Walsh and Buckleton 2005, 45). It is assumed that someone who rejects the 'voluntary' DNA test has something to hide (Prainsack 2010; Zadok, Ben-Or and Fisman 2010).

Certain groups of applicants are forced to take a parental test, as their applications will be rejected otherwise. Müller (2012, 22) states in a working paper published by the German Federal Office for Migration and Refugees that the use of DNA testing

> has created an ambiguous situation: since the burden of proof of parenthood lies with the migrant who is seeking to obtain a residence permit on the grounds of his parenthood, the … authorities can suggest the migrant to submit the results

of a voluntary DNA test to establish who the biological father is (Franßen-de la Cerda 2010, 82). Although the paternity law that was amended in 2008 prohibits public authorities from urging men to take a test to prove they are the biological father of a child, the foreigner's authorities are invoking the special burden of proof on migrants to prove all circumstances that may help them to obtain a residence permit.

Informational Self-Determination Vs Genetic Surveillance

The first DNA tests for family reunification in Germany were carried out in 1992 (Geneticist 2). Until 2010, however, they were conducted in a legal grey area. While proofs of kinship by DNA analysis were a more or less common institutional practice in this period (Federal Government of Germany 2008), they were used without any legal foundation. DNA testing for immigration purposes was first mentioned in the Genetic Diagnostics Act (Gendiagnostikgesetz), which came into force on 1 February 2010 and contains a section dealing solely with kinship DNA testing (Section 3, Paragraph 17, GenDG). The general focus of this law is on the right to informational self-determination, with the aim of protecting individuals from the abuse of their genetic information.[10]

However, for the use of DNA profiles in the context of family reunification important legal guarantees are inoperative. Firstly, as we saw in the preceding section, the right to informational self-determination in the context of family reunification is just a formal or theoretical right. In practice, it may well be the only chance a person has actually to reunite with his/her family if the documents they have provided are not deemed appropriate to prove family relations. Legal scholars do not always perceive this as a problematic restriction on self-determination and personal autonomy. In an influential commentary on the Genetic Diagnostics Act

10 The right to 'informational self-determination' is grounded in the general right of personality as stated in Article 2(1) and Article 1(1) of the Basic Law which protects the personal sphere of life, guaranteeing respect for human dignity and the right of free development of one's personality. The term was introduced by the German Federal Constitutional Court (Bundesverfassungsgericht) in a ruling relating to personal information. In its decision on 15 December 1983, it held that 'in the context of modern data processing, the protection of the individual against unlimited collection, storage, use and disclosure of his/her personal data is encompassed by the general personal rights of the Basic Constitutional Law. This basic right warrants in this respect the capacity of the individual to determine in principle the disclosure and use of his/her personal data. Limitations to this informational self-determination are allowed only in case of overriding public interest'. The Court stated that every German citizen has in principle a right to control the flow of all information relating to them. It also reasoned that the present and future conditions of data processing and collection permit a wide variety of possible abuses against which the individual has to be protected (65 BVerfGE 1, 1983).

one author stresses the voluntary character of the test, arguing that no one forces a person to apply for family reunification in the first place:

> The critique that the GenDG violates the rule of voluntariness fails to see that an application in accordance with the Passport Act and the Residence Act is completely voluntary. No one is forced to apply for his or her naturalization or family reunification. Cooperation in the procedure is the necessary consequence for those who claim rights in accordance with German law. (Kern 2012, 213)

Such a comment, which focuses exclusively on the legal argument, completely ignores the motives and the living conditions of refugees and asylum seekers; it is not their deliberate choice to seek refuge in a foreign country. Those who have fled from war zones, in particular, would mostly prefer to stay and live together with their families in their home countries. The argument that they are applying voluntarily for family reunification thus fails to take into account the often traumatic experiences of refugees and asylum seekers and the material constraints in which they find themselves.

Secondly, applicants for family reunification have no right to decide what happens to the DNA profiles once the test has been carried out. They cannot demand that their samples be destroyed, and their data might be used for criminal prosecution purposes if there is reasonable suspicion that a criminal offence has been committed by the immigrant in question (Section 3, Paragraph 17, Sentence 8, GenDG). Their profile may be stored in a DNA database in accordance with the Prüm Convention and this information may be exchanged for crime prevention purposes among all countries that signed this treaty (Toom and Prainsack 2010).[11] Therefore, applicants for family reunification may come under general suspicion of criminal activity. It has to be emphasised that this holds true not only for immigrants but also for German citizens who apply for family reunification with foreign family members. Thus, while the Genetic Diagnostics Act in its entirety strengthens the principle of informational self-determination, it denies applicants for family reunification this right. In general, the legal framework for DNA testing in family reunification cases establishes an environment of mistrust towards

---

11 The Prüm Convention is a treaty between Austria, Belgium, France, Germany, Luxembourg, The Netherlands and Spain and was signed in 2005. The participating countries agreed to render their databases searchable for the others on a hit/no hit basis – which means that another member state cannot see the actual profile but only whether the request matched a particular profile or not – to allow the exchange of DNA profiles, fingerprints and vehicle licence numbers. In 2008, the convention became a EU Council Decision (2008/615/JHA) to be applied to all EU countries. Additionally, in 2009, Iceland and Norway signed an agreement to enact certain provisions of the Prüm Convention, but as of March 2014 neither country has ratified the agreement (see http://ec.europa.eu/world/agreements/prepareCreateTreatiesWorkspace/treatiesGeneralData.do?step=0&redirect=true&treatyId=8402 [accessed 10 March 2014]).

applicants and particularly towards immigrants, and provides new means for placing them under surveillance (Aas 2006, 2011).

The use of DNA testing is considered to be an appropriate measure to prevent fraudulent uses of family reunification (European Commission 2011; Müller 2012).[12] In the Green Paper of the European Commission, the use of DNA analysis is mentioned in question 10, which deals with 'frauds, abuse, procedural issues' (European Commission 2011). It is noteworthy that DNA testing is explicitly linked to the fraudulent use of family reunification, instead of being presented as an option offered to the applicants. As a result the DNA test may rather be understood as an instrument of genetic surveillance of certain groups of people (Garner 2007). The list of blacklisted countries in Germany is quite illuminating in this respect. The Federal Foreign Office states on its website that the decision on these countries is made on the basis of an individual evaluation of the likelihood of falsified documents in the country involved. However, it is quite striking that this list almost exclusively consists of countries in sub-Saharan Africa and Central and South East Asia. Documents from countries in these regions encounter systematic mistrust, and as a consequence it is mostly applicants with black skin or families from Central and South East Asia that are requested to undergo parental testing. At the same time, citizens from other third countries (for example the USA) or nationals from other developing countries (for example in Latin America) are not usually required to provide DNA evidence (Verband binationaler Familien und Partnerschaften 2012; Immigration officers 1, 2; NGO officer 2).

One might ask if the formal and procedural argument put forward by the immigration authorities serves to target and discriminate against particular groups of applicants who will generally – regardless of their particular case – encounter more problems than other applicants in proving their identity and family relatedness. Refugee advisors and representatives of NGOs we interviewed report that applicants from countries such as Somalia, Eritrea, or Burma are almost always asked to provide DNA evidence for their family relations (Lawyers 1, 2, 4; NGO officers 1, 2). Furthermore, DNA analysis often seems to function as an additional obstacle instead of a voluntary option, particularly in the case of applications from blacklisted countries. As mentioned above, immigration officers will not trust any legal documents from applicants in these countries. However, they may well still demand that the applicants present certificates of birth and marriage, which often requires them to request these documents in their home countries in a time-consuming and costly procedure. Only after these documents are provided will the immigration authorities proceed with the application, inform the applicant that the documents cannot be accepted, and ask them to provide evidence for the biological

12 Interestingly, the head of an Aliens Department in a German city declared that even without the option of the DNA test his office is able to prevent the misuse of the right to family reunification by carefully assessing the case and verifying the documents. Nevertheless, he advocated the use of parental testing for applicants from Asia and Africa (Immigration officer 1).

family relation in the form of a parental test. In other words, the documents always have to be presented first to proceed with the administrative procedure, even if it is clear from the beginning that the authorities will not acknowledge them (Lawyers 1, 2, 4; UNHCR senior officer; NGO officer 1).

Social Vs Biological Concept of the Family

The strong focus on a biological family model in family reunification cases contrasts with the social understanding and the legal framing of the family in Germany. The routinisation of divorce and remarriage and the legal recognition of same-sex unions with the introduction of the Life Partnership Act (Lebenspartnerschaftsgesetz, LPartG) in 2001 have generated heterogeneous patterns of family structure and a diversity of new kin connections that are not necessarily based on biological ties. In recent years, several laws have come into force or been amended with the aim of emphasising the social aspects of parenthood and paternity, for example the Children's Law Reform Act (*Kindschaftsrechtsreformgesetz, KindRG*) of 1998 and the amendment of the regulations for paternity fraud. In this perspective, parenthood is not defined in terms of biological relatedness but rather as a social relation. The Federal Court of Justice (2008) argued in a judgement that this kind of socio-familial relation exists if the legal father has been shown to be responsibly looking after the child. If this is the case, the biological father cannot question the already existing social fatherhood. This line of argumentation as used by the Federal Court of Justice has been upheld by the European Court of Human Rights (*Ahrens* v. *Germany* 2012, *Kautzor* v. *Germany* 2012).[13]

Furthermore, married and unmarried couples have increasingly been treated equally in legal practice in the last 10 to 15 years, for example in tax law, inheritance law and as regards entitlement under pension schemes. It has also been made easier for unmarried couples and same-sex partners to adopt stepchildren (Paragraph 9, LPartG). These legislative steps also stress the family as a social relation. In a press release in 2009, the Federal Constitutional Court (2009) pointed out that 'biological parenthood is not prioritized over legal and social notions of the family in the jurisdiction of the Federal Constitutional Court' (see also Fehrenbacher 2009).

However, in German immigration law a different trend can be observed. The increasing use of parental testing in family reunification cases endorses an understanding of family as a biological entity. Consequently, migrants will find

13 It has to be noted, though, that the ECHR had previously also decided in favour of a biological father to allow adoption of his biological child. In this case, the child was born out of wedlock and the father was originally unaware of the pregnancy and birth of the child. When he found out, he proposed to adopt the child. The ECHR found the biological father's wish to adopt to be in the best interest of the child (*Görgülü* v. *Germany* 2004, Bainham 2008).

it difficult to enter Germany if they adhere to a different concept of family. The restriction to the nuclear family model makes it nearly impossible for larger families that consist of more than just parents and their biological underage children to reunite in Germany (European Migration Network 2010a). Relevant carers who live in the same household and relatives with whom important emotional bonds exist (for example the grandparents) almost always have to stay in the migrant's country of origin. Even foster children and adopted children are not covered by the biological family concept. According to German immigration law, no distinction shall be made between biological children and adopted children if the applicants can prove the adoption by official documentation. However, as German authorities regard official registration systems in many countries as insufficient this will lead to problems in administrative practice. Lawyers we interviewed stated that it is almost impossible to prove an adoption by legal documents, because these documents have either never been issued or are not acknowledged by the immigration authorities (Lawyers 1, 2, 4). The lawyers' statements have also been confirmed by immigration officers we spoke to (Immigration officers 1, 2). In such cases, investigations by lawyers officially accredited by the German embassies in the country of origin are required to allow for successful family reunification (Federal Government of Germany 2008). These lawyer-assisted investigations are not only difficult to initiate but also quite expensive. 'We had to pay 300 euros per document and they asked the lawyer to verify four of our documents', said an applicant from Uganda in an interview (Applicant 7).

Another problem of parental testing for family reunification results from the German authorities' current practice of asking for both a maternal lineage test and DNA kinship reports on all family members – including the putative father. The German Lawyers Society reported the case of an Afghan family where the father only learned from the DNA test in the visa application procedure that one of the children born in the marriage was not his biological child (German Bar Association 2009). Such discoveries are a heavy burden for all family members, and may result in estrangement between family members and in extreme cases in family separation instead of reunification.

The administrative practice of family unification in Germany displays a substantial legal and social difference between and contradictory treatment of native citizens and immigrants. The latter have to comply with a traditional heterosexual biological family model in order to be officially recognised as family in immigration cases. At the same time, in public opinion and in recent legislation the family is mostly defined by social dimensions for native German citizens. As social aspects such as affection and caring for each other become more and more important, the biological family model seems increasingly inadequate to capture the complexities and dynamics of contemporary family relations (Stacey 1996; Beck-Gernsheim 1998; Chambers 2012). Furthermore, the legally enforced definition of the family as biological ancestry often contrasts with concepts of family in the migrants' countries of origin. In many cultures, the term 'family' goes beyond biologically related persons by including persons to whom only social relations exist. A case in Denmark impressively illustrates this problem: 'Of all the Somalis who were subjected to DNA testing by the Danish Immigration Service from January 1997 to September 1998, 58% received a negative result. Somali community leaders responded to these findings by stating that "the concept of family is very different in [Somali] culture, and many Somalis are not aware of the Danish concept of who is a family member and thereby entitled to family reunification"' (Taitz, Weekers and Mosca 2002b, 26–27).

The problems that arise from these conflicting definitions of the family for German citizens and immigrants can be illustrated by a parental test that was carried out in a German laboratory in 2010. The application for family reunification by a man, a woman and a girl from Somalia had been turned down by the authorities earlier that year because they questioned the authenticity of the documents provided. Thus, the applicants had to resort to a DNA test to prove their family relatedness. The test result showed that neither the alleged father nor the alleged mother was biologically linked to the child. In other words, the test result demonstrated that these three persons were not a family in terms of biological relatedness. However, the staff of the laboratory were so convinced that the applicants were indeed a 'true' family, even though the result was negative, that they wrote letters to the immigration authorities arguing that the result could only be explained by an exchange of children in the hospital and that they would nevertheless advocate family reunification. It is quite remarkable that experts working in the institution that produced a test result which eliminated the possibility of a family relation according to genetic criteria still regarded the three persons to be family members. The case was reported to us by the geneticists involved in the procedure (Geneticists 3), and their argument was clearly based on a social definition of the family.[14]

These different standards for family recognition exist not only in Germany but in many host countries (see for example Degtyareva 2011). Murdock (2008, 1521) describes the situation in France as follows:

> While parenthood is considered a multi-faceted relationship for French citizens, immigrants are limited to defining the parent-child relationship as genetics-based. While French family law was modified to prevent the creation of any formal distinction between legitimate and illegitimate children, the differing standard applied to immigrants creates a new foundation for disparity. Children of immigrants who are not genetically linked to their parents are given a secondary status and the relationship is not recognized, despite the fact that the child may be with the only parent she has ever known. This dual standard implies that French families may recognize multi-faceted relationships, but without documentation, an immigrating family may only be trusted when they assert the most basic nuclear family structure.

14 At the time of the interview the applicants were still waiting for the decision of the immigration authorities.

As family policies in many host countries favour social notions of the family, there is a possibility that the gap between the legal status of native citizens and immigrants may grow, as for the latter group the focus is on biological relatedness while same-sex partnerships and patchwork concepts of family are not equally recognised.

Verifying the Family: DNA Testing as a Truth Machine

To understand the specific focus on family as a biological entity in German immigration policy and to highlight the 'German way' of using DNA evidence in family reunification cases, it is helpful to turn to the concept of 'geneticisation'. This term was coined by the Canadian sociologist Abby Lippman at the beginning of the 1990s to account for the social and cultural impact of the new genetics (Lippman 1991). Lippman employs the neologism to critically analyse a (medical) perspective that conceives of genes as a programme for the development and regulation of organisms, and regards genetics as the central conceptual model to explain human life and behaviour, health and disease, normality and deviance. Since its original formulation, the term has been used by many social scientists to designate 'a social process through which concepts, theories, social structures, and individual and social practices … are gradually changing to recognize and incorporate the explanations and rationalities of the new genetics' (Lippman 1991, 19; see also Fitzgerald 1998; Koch 2002; Ten Have 2004). While some theoretical and normative underpinnings of the concept have met with significant criticism (Novas and Rose 2000; Gibbon 2002; Rouvroy 2008), it is still useful as an analytical instrument and descriptive notion (Kollek and Lemke 2008).

The medical anthropologist Kaja Finkler has used this theoretical perspective to analyse the impact of the new genetics on family relations and kinship. Her study is based on extensive fieldwork in the USA and interviews with breast cancer patients, healthy women from families affected by breast cancer, and adopted children seeking contact to their biological parents. The results show that the ideas of her interviewees on family and kinship are dominated by a 'genetic inheritance ideology' (Finkler 2000, 10; see also Finkler, Skrzynia and Evans 2003; Konrad 2005; Edwards and Salazar 2009). Concepts of inheritance and the significance of genetic transmission informed the experiences of the individuals and their relations to their children, parents and other relatives. Finkler demonstrates that family and kin connections are framed in terms of genetic ties. In this geneticised perspective, families are defined less by voluntary bonds than by a common genetic heritage. The genetic map postulates proximity between individuals who are separated by spatial or genealogical distance. It transcends emotional alienations and conflicts between family members. Even family members who no longer see each other or who have broken off contact remain genetically close. In this perspective family ties are primarily experienced in terms of genetic heritage, and genetic bonds are seen as more fundamental and stable than family ties based on love, affection and the free will of the partners (Finkler 2000; Finkler, Skrzynia and Evans 2003).[15] The idea of a common genetic origin determines the self-image and the identity of family members. It is important, however, to take into account another aspect of geneticisation: genetics not only shapes personal identity and concepts of family and kin, but it might also conflict with existing self-concepts. In institutional contexts, genetic tests often function as a kind of 'truth machine' (Lynch, Cole and McNally 2008) that organises an epistemological field of visibilities and statements and determines what is within the realm of truth (Foucault 2000). This machinery promises to verify the 'real' sex (Dickinson et al. 2002),[16] the 'genuine' criminal offender (Lynch, Cole and McNally 2008) or the eligible asylum seeker (Tutton, Hauskeller and Sturdy 2014), and claims priority over non-genetic forms of explanation and experiences derived from everyday life (Lemke 2004).

The use of DNA analysis in the context of family reunification can accordingly be conceived of as a 'truth machine' that will 'reveal' the 'true' kin and help to distinguish them from the 'pretended' family member. As we will see (Chapter 4), DNA testing fulfils a different role in Austria where the test is used to test the social credibility of the applicants and thus functions as a lie detector. In Germany, by contrast, the use of DNA analysis in administrative decision making on family reunification promises to produce 'true' knowledge of the body and biological relatedness. It follows that personal qualifications or characteristics like reliability or authenticity do not play any important role in the administrative process. Moreover, immigration authorities in Germany place the highest degree of trust in DNA evidence when deciding whether or not a family relation exists. Verifying family bonds by DNA testing is seemingly more precise than identity papers and official documents, and claims to provide a transparent access to the body by making visible lines of belonging and networks of kinship. The idea of family verification by DNA analysis not only neglects the complexities of the technology and its internal limits.[17] It also relies on the erroneous idea of a stable

15 See also the results of the qualitative study conducted by David Armstrong, Susan Michie and Theresa Marteau. They analysed genetic counselling sessions, and investigated how the identity of the counsellees is constructed in the counselling process and how the genetic definition of family was endorsed by the counsellors at the expense of social concepts: 'The processes of exploring the family tree and mapping the genetic links therefore serve to state forcibly that genetic status was non-negotiable. There was never discussion of "who you are" as this was pre-given by the density of the genetic map: identity was located in genetic make-up' (Armstrong, Michie and Marteau 1998, 1657).

16 The practice of 'gender verification' by applying genetic knowledge and testing device in international sport serves to illuminate the operations of this 'truth machine'. This control mechanism resulted in a number of women with seemingly 'male attributes' (for example females with androgynous insensitivity, XY gonadal dysgenesis) who were excluded from competition (Ljungqvist and Simpson 1992; Simpson et al. 2000; Dickinson et al. 2002).

17 While parental testing is a very precise way to determine biological relatedness, the technology also has some limits: for example, mutations may occur that lead to an

and unchanging body (Mol 2002; Lynch, Cole and McNally 2008, 191) that can be addressed as the basis for identification, and supports the phantasm of fictitious control by technological means that allows for more safety and efficiency. However, the search for identification and trust is the other side of a culture of mistrust and imagined danger by 'unidentified subjects' 'whose minds cannot be trusted but whose bodies do not lie' (Aas 2006, 156).

This ambiguity of parental testing for family reunification in Germany and many other countries also points to the contradictory implications of biotechnological innovations for the concept of the family. On the one hand, reproductive technologies such as in vitro fertilisation or egg donation allow for a detachment of social parenthood from biological relatedness. Some social scientists have argued that parenthood can be disassembled into its integral parts – in analogy to a construction kit – and recombined in a completely flexible way (Beck 1990). On the other hand, the same technologies are used to reproduce and reinstate the idea of an unchangeable body and 'natural' concept of family (Weigel 2002). As a consequence, biological notions of family and kinship are strengthened further. Here we note an interesting paradox: while genetic and reproductive technologies undermine the idea that family relations and parenthood are natural relationships founded on biology, they are simultaneously being employed to reaffirm and re-establish this idea. Apparently, the decomposition of 'natural' family ties and the reconfiguration of family concepts by the use of genetic technologies come at the price of a naturalisation of society (Franklin 2000).

Conclusion

In this chapter we have discussed some implications that result from the use of DNA testing in the context of family reunification in Germany. While it is certainly a legitimate interest of the German political authorities, and generally of all nation states, to regulate immigration and to prevent fraudulent family reunification, the trend to increasingly rely on DNA testing raises three important concerns. Firstly, we pointed out that there is considerable tension between the official and legal presentation of DNA testing as an additional and voluntary choice in administrative decision making on family reunification and the effective constraint to take up this option – since the application will be rejected otherwise. Certain groups of applicants – those who originate in the blacklisted countries – will be more or less obliged to undergo parental testing in order to provide evidence considered satisfactory by the German authorities. Secondly, the German regime of processing and evaluating DNA evidence in applications on family reunification promotes the idea of a scientific and safe basis of family relations that 'trumps' any other kind of evidence to document family relations (from interview material to identity papers).

exclusion of maternity or paternity even though a biological relation exists (Dawid, Mortera and Pascali 2001; Morling et al. 2002; Karlsson et al. 2007; Mansuet-Lupo et al. 2009).

This legal and administrative practice fails to adequately take into account privacy concerns and the right to informational self-determination concerning genetic data. Thirdly, the use of DNA tests in the context of family reunification nurtures the idea of a biological family and tends to devalue alternative conceptions of family as less important and secondary. The requirement of a genetic link between family members contrasts with family recognition policies and legislation in Germany, and in many other host countries, which emphasise social rather than genetic ties.

To be clear: the argument presented here is not directed against the use of DNA kinship tests in family reunification procedures as such; nor is it our intention to argue for a ban on their use in the context of immigration. Quite on the contrary: under certain conditions, these tests may well be a helpful tool for both applicants and immigration authorities, for example if no official documentation and identity papers are available and the application would be rejected otherwise.

However, the findings of this empirical study indicate that a more balanced approach is needed in German immigration policy, one that acknowledges the complexities and limits of DNA kinship testing as a way of verifying family relatedness. It is necessary to reconsider administrative decision making to take into account privacy concerns, issues of data protection, and the diversity of family arrangements in contemporary societies. The institutional focus on DNA testing as a way of proving family relations and the mobilisation of a biological concept of family which consequently marginalises social aspects of family life produce new constraints for immigrants. They establish double standards as applied to native citizens and immigrants in respect of their legal and political rights, and promote a culture of mistrust and fear.

Chapter 3
Finland: Securing Human Rights, Suspecting Fraud[1]

Anna-Maria Tapaninen and Ilpo Helén

Introduction

Finland was a pioneer in the adoption and application of DNA testing for family reunification. It was one of the first countries to incorporate the regulation of such testing in a law and to establish it as a standard procedure in the administration of immigration; this happened in the late 1990s and early 2000s. In addition, the process of law-making and its implementation has been rather transparent, and the relevant documents are quite easily accessible in Finland. The Finnish system is highly centralised, as the Finnish Immigration Service is responsible for the processing of all applications for residence permits and asylum and also for instigating DNA tests. For these reasons, Finland has figured as an exemplary case in the research literature (La Spina 2012; Taitz, Weekers and Mosca 2002b).

In this chapter, we situate DNA testing for verification of family ties in the context of the late 1990s and early 2000s Finnish migration regime. The subject of our study is twofold. First, we are interested in the framework of immigration policy in which the technology of DNA analysis was initiated and established and in which it is deployed today. Second, we analyse the role and impact of DNA testing in policy and administrative practices. Our study is guided by three questions: What are the problems of immigration management to which DNA analysis has been thought to provide solutions? How does the testing 'solve' the problems? What sort of new problems, tensions and contestation does the use of DNA testing of family ties engender? By analysing policy documents and expert interviews (see below), we disclose the *practical reasoning* of Finnish immigration authorities, policymakers and experts: how do they define the problems, what role do they allot to DNA testing in policy and administrative practice, and what do they expect from the biological 'tool'? From a critical social research perspective, we place the views and reasoning of the experts in the context not only of legislation and policymaking but also of concrete administrative work and decision making.

1 A version of this chapter has been published in *Nordic Journal of Migration Research*, 3(3), June 2013, by Walter de Gruyter, all rights reserved. © Ilpo Helén and Anna-Maria Tapaninen, 2013 (Helén and Tapaninen 2013). For the purpose of the present volume, the text has been expanded, substantially revised and updated.

Through multiple contextualisations, we present a nuanced picture of today's immigration management in Finland.

We begin our chapter by describing the immigration regime in Finland in the 1990s as the context for initiating DNA testing. Then, we proceed to analyse the implementation of the testing procedure from the pilot project in 1996 to legal and administrative normalisation of the use of DNA proof for family reunification in the early 2000s. For our analysis, we use contemporary policy documents and interviews in which experts and policymakers present how they understand the beginnings and objectives of DNA testing in light of current problems and policy. Our main finding is that the use of DNA analysis for family reunification has been framed by a continuous coupling of two policy rationales, securing human rights and combating fraud, during two decades of its deployment for immigration management in Finland. We emphasise that despite obvious tensions between the two rationales, they are not mutually exclusive but exist in parallel, providing a sort of 'double track' for reasoning about and justification of the use of DNA analysis within the Finnish migration regime. In our conclusion, we discuss recent trends in the use of DNA analysis for family reunification in Finland.

Data and Methods

Our study is based on policy and administrative documents from the mid-1990s to 2010s and on 25 interviews with experts in immigration administration, laboratories, NGOs and law firms. The research material also included discussions in seminars in which experts, immigration authorities, and politicians discussed family reunification, email correspondence, and communications by telephone with additional experts. Some of the interviewees were contacted several times in order to update and clarify information concerning changing administrative practices. In addition to the official documents related to legislation and administrative procedure we received unpublished documents from the interviewees, including internal reports related to the pilot project on DNA testing.

In many ways, finding informants, conducting interviews and gathering material was akin to fieldwork. In this sense, our research follows the methodological approaches of grounded theory and ethnographic research (for example Birks and Mills 2011; Clarke 2005; Hammersley and Atkinson 2007). Embedded in and directed by our data, our analysis attempts to capture the use of DNA tests for family reunification as it is understood, conceptualised and reflected on by those involved in actual practices within Finnish immigration management. We consider our study a process in which a strictly structured approach would not yield valuable results. Instead, the interviews and encounters are illuminating snapshots of evolving narratives within the field. In what follows, we do not specify the data quantitatively or analyse the individual institutions in question. Nor do we attempt to decipher the attitudes or opinions of single informants or stakeholders.[2] Rather, we understand our data in the context of dialogically unfolding discussions in a dynamic and contested field in which the story of the beginning of DNA testing is narrated and the rationales and justifications for testing are formulated by the authorities and experts involved.

The Immigration Regime of the 1990s

The introduction of biotechnological methods in immigration administration was related to rapidly increasing immigration to Finland during the 1990s. 1990 was a watershed year in post-WW II Finland. After the collapse of the Soviet Union Finland's eastern border, which had been all but closed during the cold war, was 'opened'. Refugees, especially Somalis who were fleeing from their war-ridden homeland, now began to cross this border. Even though the number of asylum seekers was small compared to the other Nordic countries, mobility from the global South was an unforeseen phenomenon in a peripheral country with a long history of labour emigration rather than *Gastarbeiter* immigration (by comparison with Germany, Austria and Sweden) or a postcolonial immigration regime (as in the UK and France). The borders were also opened in relation to international conventions and regulations. Finland joined the European Council in 1989, ratified the European Convention of Human Rights in 1990, and joined the European Union in 1995. During the same years, Finnish Ingrians from the former Soviet Union also began to cross the border and were received as returnees.[3] Both groups, the Somalis and the Ingrians, received a lot of media attention and their presence instigated legislative amendments, even though there were great differences between them as newcomers.[4]

---

2 It is noteworthy that not only do relevant authorities and experts communicate and collaborate with each other, but individual experts, especially lawyers, can also change their institutional affiliation and workplaces between, for example, migration administration, NGOs, independent legal firms, UNHCR, diplomatic missions, and the courts of law. Therefore, the interviewees typically do not represent only one perspective. Rather, the answers given seem to reflect a dialogical constitution of official and unofficial narratives. This said, we do not aim to deny that there are evident tensions between the different stakeholders. Because our contacts were all working within a restricted field, it has been a particularly challenging task to avoid disclosing the identities of the informants.

3 The Ingrians are the Finnish population of Ingria (now the central part of Leningrad Oblast of Russia). They are descended from Lutheran Finnish immigrants from Savonia and Karelia, who were introduced to the area in the seventeenth century when both Finland and Ingria belonged to the Swedish Empire (see, for example, Teinonen and Virtanen 2000). The history of the movements of Ingrians across the border is quite symptomatic of major political changes.

4 President Koivisto's views epitomised a more common political sentiment at that time. He personally demanded that the Finnish Ingrians be received as returnees on the

The decade began with media headlines about a 'mass migration' of refugees from Somalia arriving by train from Moscow (Aallas 1991; Hautaniemi 2004; Leitzinger 2010; Lepola 2000). At the end of 1990, 250 Somalis arrived within two days, which 'became the number one piece of domestic news' (Lepola 2000, 9). President Koivisto reacted to this forcefully and demanded that 'the Somali flood' be stopped (Lepola 2000, 80). Even though the large number of daily border crossings was emphasised, the total numbers were relatively small: by the end of 1990, fewer than 1500 Somalis had applied for asylum (Leitzinger 2010: 85). The public dramatisation of immigration testified to the unpreparedness and almost xenophobic fear of many Finns when encountering a new phenomenon. For public authorities, managing the asylum seekers was problematic, and they repeatedly emphasised that the asylum seekers did not usually have proper identity documents or visas. The Finnish Ingrians were also suspected of being 'fake Ingrians' (Leitzinger 2010, 80–85). However, the documents of the former Soviet Union and Russia were seen as more reliable. In contrast to Somali nationals, the Ingrians' applications were considered easier to process since, as an immigration official said: 'We speak the same language of bureaucracy' (Official 1). The language of asylum seekers appeared more complicated and suspect to the immigration authorities. The documents they presented were dismissed as forged, and the narrative evidence of interviews was considered fraudulent.

Soon after the first asylum seekers from Somalia had entered Finland, many Somalis filed applications for residence permits for their family members. These required new forms of regulation. In Finland, family-related migration was connected almost exclusively with refugees in public discussions and in comments by policymakers and authorities (for example UVI 2005). Today the scene is more heterogeneous, and the number of applications and residence permits granted to family members of recipients of international protection has declined dramatically since 2011, amounting to only 6 per cent of all residence permits based on family ties in 2013 (Maahanmuuttovirasto 2014). However, asylum seekers and refugees and their 'complex' family ties are regularly highlighted in discussions of family migration and immigration in general.

Asylum seekers did not always have identity documents or certificates of birth, marriage, divorce, adoption or death. Official documents may either have been lost when fleeing, or the certificates of family ties and the related registers may not have existed in the country of origin. At the beginning of the 1990s, these questions were tackled in heterogeneous ways which led to administrative, legal and technical reforms. The Directorate of Immigration set a goal of more efficient decision making in 1996, including the improvement of interview methods and the introduction of DNA testing. As an official said, 'an idea was born that in order to get closer to the truth let's instigate a project and investigate the potentials of

basis of an amendment to the Aliens Act (which was in force until 2011). At the same time, he was critical and/or indifferent when commenting on the arrival of the Somalis (Aallas 1991; Lepola 2000).

the use of DNA' (Official 9). DNA testing was seen as an appropriate solution, especially to complications in the processing of applications submitted by Somali refugees. Hence, DNA testing was justified as an efficient tool in the exceptional case of the Somalis; however, it quite soon became a relatively uncontested routine procedure.[5]

Initiation of DNA Testing

In the following, we describe the process that led to the amendments of the Aliens Act in 2000 and to the established administrative procedures of DNA testing for family reunification as it is displayed in official documents, unpublished primary sources related to law–making, and the expert interviews.[6] The Finnish documents emphasise that the 'Somali problem' made the biotechnological method topical in the late 1990s. A particular difficulty in the processing of applications by Somalis was highlighted: they had mostly received negative decisions. For example, 652 out of 775 applications for family reunification (84 per cent) were rejected in 1995 (UVT 1996a). Discussions among immigration officials and experts aimed at swift reconsideration of the rejected applications for family reunification because of a considerable number of underage Somali applicants, half of whose applications had been rejected (UVT 1996a; UVI 1996). It was also pointed out that some of these children had waited for their family members for as long as five years (UVT 1997). Another reason why reform, including DNA testing, was seen to be necessary was the fact that at that time the Aliens Act did not provide any right to appeal: 'You needed to be creative and work out the best possible solutions' (Official 9). Furthermore, the refugee office of the Ministry of Social Affairs and Health, the Office of the Ombudsman for Foreigners, and the Refugee Advice Centre had been contacted by several dozen Somalis whose family reunification applications had been rejected because of a lack of valid documents and inconsistencies in interviews (Kokkarinen n.d.; Official 20). In this situation, the use of DNA analysis was seen to be justified both by general principles of

5 Somalis have also been singled out in the context of DNA analysis in other countries. Difficulty in processing their applications in Hungary was the background to the Ad Hoc query on DNA testing by European Migration Network (EMN 2009). In Denmark, DNA testing was first used in the cases of Somalis and only later became a more extensive practice (HE 88/1999). They were also targeted in the United States (Villiers 2010).

6 Our main official sources are the Annual Reports of the Ombudsman of Foreigners (UVT 1996a, 1997), the report of the fact-finding trip by the Office of the Ombudsman of Foreigners (UVT 1996b), the memorandum Letter of Directorate of Immigration to the Ministry of the Interior (UVI 1996), the internal report of the pilot project (Kokkarinen n.d.), the Statement of the Ministry of Justice to the Ministry of the Interior (OM 1999), Government Proposal (HE 88/1999), statements given to the Administrative Committee by authorities (HaVL 1999a, 1999b, 1999c. 1999d, 1999e), and the amendments (SDK 114/2000) to the Aliens Act (SDK 378/1991).

human rights and also by actual cases that were waiting to be resolved. In fact, the initiative for DNA testing came from authorities and NGOs that had been in direct contact with the applicants. Once again, the Somalis were singled out: they were the most numerous group of applicants for family reunification, their applications were difficult to process, and they had received predominantly negative decisions.[7]

Representatives of immigration administration, NGOs and laboratories discussed the above issues in several meetings in 1995 and 1996 (UVI 1996; Official 20).[8] As a result, a 'fact-finding' trip was made to Kenya and Ethiopia in 1996 to investigate the possibility of and facilities for DNA testing. Information was gathered on experiences of and views on DNA testing from missions and international organisations.[9] The parties unanimously suggested that the possibility of using DNA tests in family reunification procedures should be studied in a pilot project (UVI 1996). The project as a whole was clearly seen as urgent by the authorities and experts involved, and respect for human rights, especially the rights of children, was the main motivation and justification for the immediate initiation of DNA testing trials (Kokkarinen n.d.; UVT 1996; UVI 1996; HaVL 1999d; HE 88/1999; Official 6). The Finnish experts also informed the Somalis of the principles of conducting DNA testing for family ties. According to an official involved in the pilot project:

> The project was made known, [there were] discussions with the Somali community, organization, I recall. They were invited to meetings and they were told. If I recall correctly, the elders of organizations were told and they spread the word. We explained what the DNA system was and that is leads to certainty. There were no moral problems or discussions that it would be ethically wrong. … I remember that an elderly person said that as they cannot get documents, passports, *let the DNA test be their passport*! (Official 9)

It seems that in an encompassing world of uncertainties and doubt many of the parties involved appreciated exact knowledge, and the applicants were said to

7 In most cases, the reasons for rejections were related – rather puzzlingly – to age and to the fact that the family ties had ceased to exist (Kokkarinen, n.d.). This is one indication of the fact that the applicants were said to offer seemingly precise information to the officials for the sake of precision, which can work against their interest (NGO 8). The detailed information offered can just as well be interpreted as a fraudulent. The quest for reliable information has been a central tenet of immigration administration ever since.

8 Directorate of Immigration, Refugee Office of the Ministry of Social Affairs, Office of the Ombudsman for Foreigners, Refugee Advice Centre, Department of Forensic Medicine at the University of Helsinki, and National Public Health Institute.

9 The missions of Finland, Sweden, Denmark, Netherlands, the United Kingdom and Canada were included, as well as the UNHCR (United Nations High Commissioner for Refugees), UNICEF (The United Nations Children's Fund), UNDP (United Nations Development Programme), ICRC (International Committee of the Red Cross), and IOM (International Organization for Migration).

have rarely refused the test (Official 4; NGO 19; Finnish Immigration Service 2008; European Migration Network 2009). However, in the first year of DNA testing 47 families were offered the test, and 10 of them (over 20 per cent) refused (UVI 2001).

The pilot project was conducted between 1996–1999. The cases were selected from applicants who had received negative decisions due to missing documentation and inconsistencies in interviews. In the first phase, 43 persons in Finland (including 37 minors) and 291 family members in Ethiopia and Kenya were tested. A meticulous report on the results of the pilot project also set out the reasons why some people had not shown up for the test: death, disappearance, divorce, returning to Somalia or leaving for another country while waiting for the decision. Furthermore, the report gave details of the few cases in which DNA analysis did not confirm the claimed family ties and led to exclusions. The main conclusion was that that 'test results showed almost without exception that the family members claimed in the application were real family members' (Kokkarinen n.d.). Despite occasional inconsistencies, the test proved to be a ticket to Finland for nearly all the people tested during the pilot project (Kokkarinen n.d.; HE 88/1999, 2; HaVL 1999a, 1999d).

For politicians and officials in immigration administration, DNA testing was justified because it provided precise and reliable knowledge of family ties. By reducing the complexities of family ties to the genetic relationship, many complications were solved (Officials 1, 9). However, it was rather common in official statements to make a remark that 'true families' may include, for example, adopted children, and therefore the family cannot be defined solely by biological relatedness despite the use of DNA analysis (for example HaVL 1999d, 1999e; HE 88/1999, 3). By comparison with Germany, it seems that the Finnish immigration authorities emphasise more the social aspects of the 'family' in their consideration of DNA testing for family reunification (see Chapter 2).

From the beginning, the approach was clearly legalistic. The authorities and legislators thought that DNA testing should be regulated by legislation, especially because the basic rights of personal life, liberty, integrity, and privacy are at stake when DNA test is applied for a non-medical purpose in the family reunification procedure (Official 6; OM 1999). Legal regulation was established as the Aliens Act (SDK 378/1991) was amended (SDK 114/2000) with two sections on DNA testing.

Ethical problems were discussed in the pilot project report (Kokkarinen n.d.) and in the government proposal (HE 88/1999). The report stated that DNA testing 'cannot become the norm' and that 'it cannot be a tool used only for one ethnic group', noting that DNA testing had been linked solely to Somali refugees in the media. It also suggested that the applicants should have the right to demand testing. Further, it was emphasised that the test should be absolutely voluntary and refusing the test should not be a ground for a negative decision (Kokkarinen n.d.; HE 88/1999).

Regarding DNA testing as such, the documents do not mention ethical controversies. However, a forensic expert observed that destroying the sample and data, which was eventually required by Section 65(3) of the Aliens Act (SDK 301/2004), is an indication of the fact that genetic knowledge was seen as frightening by policymakers and legislators: 'Horrible, horrible, horrible; genes, genes, genes' (Geneticist 7). The Department of Forensic Medicine argued unsuccessfully against the definite requirement for the destruction of the data obtained from the tests (HaVL 1999c). All in all, it seems that a consensual view prevailed in Finnish policymaking and overrode potential conflicts about the ethics of DNA testing.

The statement by the Ministry of Justice to the Ministry of the Interior (OM 1999) emphasised basic rights as laid down in the new Constitution of Finland (SDK 731/1999), especially the right to life, personal liberty and integrity, and the right to privacy. As regards informed consent, the statement acknowledged that the applicant's consent was conditioned by the fact that his or her refusal to take the test would probably lead to the rejection of the application (OM 1999, 4). The scope of the test was another question raised. It was considered the final option after all other investigations had been deemed inadequate. The use of DNA test should be limited to cases where it could probably lead to positive results, a position that was also articulated in the Aliens Act (SDK 378/1991) and later incorporated in Article 65 (1) of the current Aliens Act (SDK 301/2004): 'DNA tests should be done when it is evident on the basis of other investigations that the alleged family tie does exist.'[10] In its statement, the Refugee Advice Centre extended the potential uses and emphasised that all applicants should be offered the right to DNA testing free of charge when other evidence was not considered adequate (HaVL 1999b). From the perspective of civic organisations working with the asylum seekers, this point of view remains relevant today:

> Undoubtedly, the applicants have a right to DNA testing. I do not recall any cases in which DNA testing would have worked against the applicant. So we are in favour of testing if it will provide adequate information for decision-making; DNA testing should be carried out in all cases in which evidence of the family tie is required and documentary proofs are not available. (NGO 24)

The emphasis on human rights was central in the rationale that directed the initiation and establishment of DNA testing for family reunification in Finland. Authorities and experts tend to consider DNA analysis as a right, 'the last resort' offered to applicants without proper documents, which also guarantees the principle of non-discrimination (HE 88/1991, 3; HaVL 1999a; OM 1999). For example, a geneticist and forensic scientist involved in the pilot project summed

10 This was originally Section 18e of the Aliens Act of 1991 (SDK 114/1991) as amended by SDK 114/2000. Hereafter, all references to the Aliens Act refer to the current one only because the provisions on DNA testing are identical to the original provisions.

up the findings by saying that the project 'proved that nobody had lied, they were genuine biological families' (Geneticist 7). In the eyes of this scientist, such a result confirmed the view that the main purpose of the 'truth' provided by DNA testing is to secure the human rights of the persons tested. In a parallel way, the rationale of human rights is central in numerous Finnish administrative and policy documents of the late 1990s and early 2000s (for example UVT 1996; UVI 1996; HaVL 1999d; HE 88/1999; OM 1999), as they see verification of family ties using a technique from molecular biology as a way of affirming human rights and supporting applicants' claims:

> The Directorate of Immigration considers the legislative proposal on DNA testing appropriate. With the help of DNA analysis, a correct and just result may be reached even in unclear cases. Applicants who cannot present documents to prove the family ties due to a chaotic situation in their home countries may participate in DNA testing, which verifies biological family ties, and thus have an equal opportunity to provide sufficient evidence of their family relations. … The legislative proposal touches especially upon underage children and their rights. The currently proposed amendment in law is congruent with the Convention on the Rights of the Child and advances its actualization. (HaVL 1999a)

The Law and an Established Procedure

In Finland, DNA testing was incorporated into the Aliens Act in 2000 (SDK 114/2000) and the European Council Directive on the right to family reunification of third country nationals (2003/86/EC; see Introduction) was fully transposed into national legislation by 2006 (SDK 301/2004). The legislation and policy are in some respects more liberal in Finland than the provisions of the Directive. The Aliens Act is in line with the Finnish concept of family, and the definition of the family mainly covers the members of the nuclear family. However, the legislation is relatively inclusive in some respects. For example, the definition of a spouse is quite liberal. According to the Aliens Act:

> Persons living continuously in a marriage-like relationship within the same household regardless of their sex are comparable to a married couple. The requirement is that they have lived together for at least two years. This is not required if the persons have a child in their joint custody or if there are other weighty reasons for it. (SDK 114/2000, § 37(2))

Thus unmarried couples, registered partnerships and same-sex unions are included in the legal formulation of marriage-like relationships, which makes legal framings for family reunification more inclusive in Finland than in Germany (see Chapter 2). It is also noteworthy that marriage-like relationships as defined by law can be verified through DNA analysis of parentage if there is no official documentation on cohabitation or custody. This is one of the potential extensions of the use of DNA analysis, which implies that genetic relatedness of a child to his or her parents constitutes, in a legal sense, a family or a household.

Pursuant to Article 37 (1) of the Aliens Act, underage unmarried children 'over whom the person residing in Finland or his or her spouse had guardianship' are considered family members. This legal definition does not differentiate between 'natural' and 'social' filiation, that is to say between biological children, adopted children and stepchildren. Article 37(3) further states that a child (foster child) who is under 'his or her parents' or guardians' *de facto* care and custody … shall be treated as a child under subsection 1, if there is reliable evidence on the death or disappearance of the child's previous parents or guardians'. Often the relationship between the child and the actual guardian cannot be verified by official documents of adoption. In such a case the fosterage relation has to be investigated in interviews and by DNA analysis, too, if the guardian is biologically related to the child in question, for example as a grandparent, an aunt or an uncle. Moreover, unaccompanied minors can be sponsors for reunification with their minor siblings under Article 52 (3) of the Aliens Act, which also states that '[a] requirement for issuing a residence permit is that the children and their siblings have lived together and that their parents are no longer alive or the parents' whereabouts are unknown'. This right to apply for family reunification for minor siblings is not available in either Austria or Germany.

The amendments of the Aliens Act in 2000 (SDK 114/2000) provided the legal basis for the administrative procedures of DNA testing for family reunification. As these provisions have not been amended since, these procedures have been guided by the same principles for the past 15 years even though Finnish migration policy has changed considerably in that time. Cooperation and division of labour between the authorities are defined in the Section 66 of the current Aliens Act (SDK 301/2004). The Finnish Immigration Service (former Directorate of Immigration) coordinates the information processing and makes the decision. In Finland, applicants without sufficient documentary proof are interviewed by the police who also monitor the taking of the DNA samples. Abroad, the personnel of the Finnish Diplomatic Missions, sometimes with the help of officials of the Finnish Immigration Service, conduct the interviews and carry out the sample taking and its delivery. The Finnish Immigration Service requests the tests from the Hjelt Institute (former Department of Forensic Medicine) laboratory at the University of Helsinki. The authorities involved – officials of the Finnish Immigration Service and of the diplomatic missions, the police, forensic experts, and officials of municipalities – exchange information and are regularly in contact with each other even though their roles in decision making are clearly differentiated. It is noteworthy that the Hjelt Institute laboratory sends the DNA analysis report directly to the Finnish Immigration Service and not to the applicant; this practice is unlike information management in the German procedure, in which the laboratory passes the results of DNA analysis to the applicants themselves who are supposed to present them as evidence to the immigration authorities (see Chapter 2).

The procedures of DNA testing are defined in the provisions of the Aliens Act:[11]

> The Finnish Immigration Service provides the applicant or the sponsor with an opportunity to prove their biological kinship with DNA analysis paid for from State funds if no other adequate evidence of family ties based on biological kinship is available and it is possible to obtain material evidence of the family ties through DNA analysis. (SDK 301/2004, Section 65 (1))

The applicants do not have to meet the cost of the test unless 'they have deliberately given false information, as a result of which the person and the family member indicated by him or her have been ordered to take the DNA test' (SDK 301/2004, Section 65 (3)). This principle of monetary compensation resembles the Austrian practice of dealing with alleged frauds through monetary sanctions (see Chapter 4); however, according to an immigration official it has never been applied (Official 4). Furthermore, the idea of DNA testing as a voluntary option offered to the applicant is stressed as Section 65 (2) of the Act demands 'free written consent based on information and free will'. Information on the purposes and nature of DNA testing is provided to the applicants in a Fact Sheet available in five languages, as a printed form and on the website of the Finnish Immigration Service (Finnish Immigration Service 2014). The Fact Sheet is also read out, translated and explained by the interpreters if necessary. From the human rights perspective it is also important that, under Section 65 (2), 'the results may not be used for any purposes other than establishing the family ties required for issuing a residence permit in cases as specified in the person's consent' (SDK 301/2004, Section 65 (2)).

Contrary to the justifications and intentions stated in the process of amending the legislation in Finland, DNA testing has evidently become routine (Official 4; NGO 14; NGO 19; Hautaniemi 2007). This can be seen also in the number of tests taken in relation to the number of applications (see Figure 1). According to the available statistics the number of tests peaked in 2010, when 752 tests were conducted (European Migration Network 2012b) and the number of applications was in excess of 10,000 (Finnish Immigration Service 2011). Since then the number of applications submitted by family members of recipients of international protection, especially by Somalis, has declined drastically, but the frequency of testing has not decreased at a similar pace. It appears that DNA testing is still a significant tool in the processing of applications by the targeted populations. It is considered that currently Somalis find it very difficult, even impossible, to get a positive decision without verification of their family ties by the DNA test (Official 4; NGO 14; NGO 19).

---

11 Another law regulating DNA analysis, the laws on forensic genetic paternity examination (SDK 378/2005) and the related decree (SDK 755/2005) specify the procedure in detail: the two authorised laboratories, legal rights, technical details and even the maximum prices. The legislation and the procedures have not changed, and even the prices are still the same, approximately 250 euros per person.

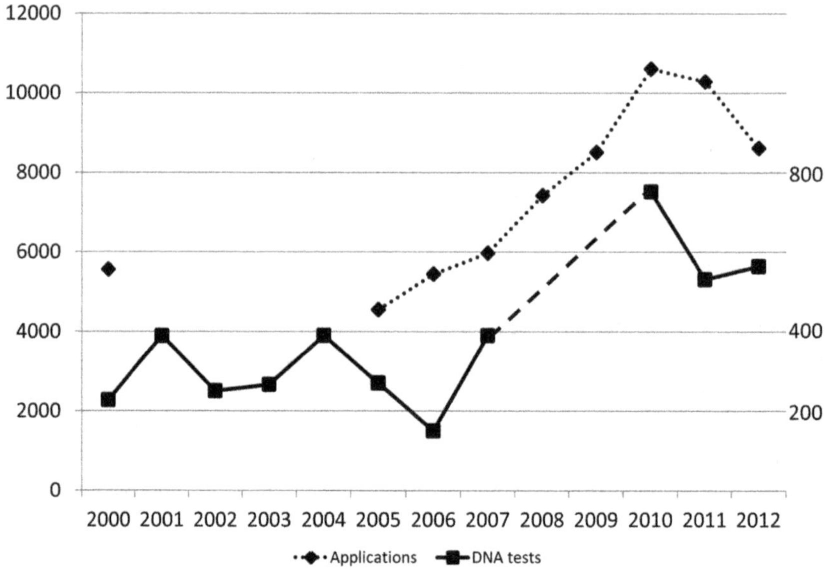

Figure 3.1 Applicants for family reunification by immigrants under international protection and the number of DNA tests for family reunification in Finland, to 2012

Sources: Finnish Immigration Service 2008; Maahanmuuttovirasto 2011, 2014; European Migration Network 2009, 2012b; Statistics Finland 2012; Official 4[12]

Fighting Fraud

The human rights perspective was dominant in the discussions when DNA testing was initiated and established in Finland in the late 1990s (see above). Despite this rationale, administrative documents and expert interviews clearly demonstrate that a general assumption of untruthfulness of applicants for family reunification has been a central motivation and justification of DNA parental testing from the very beginning. 'As wrong families and wrong relatives have come to Finland either accidentally or on purpose, the introduction of DNA analysis has been suggested', reported the Ombudsman for Foreigners in 1996 (UVT 1997). The Government Proposal (HE 88/1999, 6) stated that residence permits were granted to 120 persons 'who were not family members' in 1992–1997, and 'because of the nature of the issue, the extent of the phenomenon cannot be known'. By giving

12 The information on DNA tests is not complete. The number for 2002 is an estimate, and the tests conducted in December 2007 are not included in this figure; the total number in 2007 has been estimated to have reached 480–500. Information about the number of DNA tests in 2008 and 2009 is not available.

exact numbers while referring to uncertainty about the scale of the problem, the image of an uncontrollable threat is evoked. It was also noted that with DNA testing, only 'genuine family members' would be given a residence permit. Thus the Finnish immigration authorities and policy makers see the technique of DNA analysis of family ties as an effective tool for combating fraud. As the report of the pilot project on DNA testing (Kokkarinen n.d.) stated: 'It has been advocated as a way of eliminating misdemeanours in family reunifications.'

The administrative documents and the officials' interviews are full of references to fraud. Even in the policy documents of the late 1990s and early 2000s, false documents of identity and contradictory answers by the applicants are highlighted as problems encountered when identifying the newcomers and their alleged relatives. In general, the applicants without valid documents proving their identity, marriage or parenthood are framed through the problem of fraud, and the lack of documents merges with forged ones in the eyes of the immigration authorities. In interviews, officials tend to refer to misdemeanours in family reunification rather vaguely and they are often cautious and non-specific when questioned about actual cases of fraudulent applications. It appears that a sort of generalised suspicion prevails among the Finnish immigration authorities, and therefore they consider strong indications of fraudulent intention in individual cases secondary: applicants of certain nationalities – especially Somalis, but also Iraqis and, to a lesser extent, some African nationalities (European Migration Network 2009) – are a priori considered unidentifiable and thus suspect, and potentially subjected to DNA testing.

Furthermore, the 'multiplier effect' (Honahan 2009) of fraudulent family reunification is emphasised by the Finnish authorities. A statement by Heikki Taskinen, the director of the Immigration Unit of the Finnish Immigration Service until 2012, indicates this clearly:

> In some applications for family reunification misdemeanours began to appear. These misdemeanours were the following: some Somalis applied for a residence permit on the basis of family ties for people who were not really their family members. After getting into Finland these sooner or later reported that they belonged to another family but then, in turn, applied for a residence permits for others to whom they were spuriously related. (Taskinen cited in *Leitzinger* 2010, 86)

This tendency is frequently noted by officials and other experts as related to the problem of human trafficking. False claims about parentage are seen as problematic regarding the rights of children and as potentially leading to the neglect and even abuse of minors (Official 4; Lawyer 5;[13] Official 6; Lawyer 11). In fact, this was one of the grounds for setting up the pilot project:

---

13 Informants referred to as 'lawyers' are experts providing juridical help to the applicants.

> Originally, if somebody said that this person was a relative they were taken completely at face value; it was believed that they were telling the truth. Then, situations appeared in which we noticed that some or all of the children were not the children of that particular family, and these children began to apply for the members of their own families. Or, in worse cases, this was not even revealed and the children may have ended up in dubious circumstances because blood ties or kin relations did not exist. … So, we considered that DNA testing was good also from the point of view of the child's interest. (Official 6)

The rationale of suspicion, based on the argument that 'Finns shall not be deceived' (NGO 14), leads immigration officials to emphasise that 'we have to get the true families and not the fake ones' (Official 9). To reach this objective, they welcome the precision and objectivity of DNA analysis. It is seen as useful for revealing fraudulent applications and making the right decisions on 'truthful' applications case by case, not least because focusing on verification of genetic relatedness reduces the complexity of the cases. In addition, immigration authorities tend to applaud the indirect, self-selective consequences of testing: 'Awareness of the availability of DNA testing also forestalls applications submitted on false grounds', said Matti Saarela, Director General of the Directorate of Immigration in 2000 (UVI 2001; see also HE 88/1999). The authorities consider DNA testing an efficient means of fighting fraud *beforehand*, because the potential applicants learn about the purposes and results of the test and are therefore unlikely to file fraudulent applications:

> Of course one could say that tests were not taken by the people who possibly *knew* that the result would not be positive. And that's how it has been ever since. Only by accident has it turned out that people did not know that they weren't [biological relatives]. (Official 9)

A tendency to construct a narrative of progressive success in combating fraud seems to be a significant feature of the administrative discourse (Officials 1, 6, 9). The findings of the pilot project did not confirm this, however. Failures to show up for testing were explained by disappearances and deaths, and the unexpected test results in terms of family ties in apparently polygamous families (Kokkarinen n.d.; see also UVI 2001). However, the immigration authorities may ignore such findings by relying on their presupposition of omnipresent fraud. An official of the Finnish Immigration Service said that:

> Initially, when there were fewer applications and before DNA testing, claims that 'this is a biological child' were believed and families were considered authentic families … But when DNA testing began, there were a lot of exclusions at the beginning, that is, it turned out that some of the children were not the children of those who claimed to be their parents. Thereafter these cases have become more and more infrequent, because the applicants knew that they would be tested, and

> *there is a pre-emptive effect which was actually the initial purpose.* (Official 1, emphasis by the authors)

The Finnish immigration authorities are quite confident that applicants have enough knowledge to understand the procedure and its consequences, and that they are capable of adopting 'alternative tactics' due to this knowledge. When asked if people have agreed to be tested without knowing what it means, an official replied: 'No, they have not gone there in ignorance but there were … as happens in life, all kinds of … something may have been revealed in some cases. But people have *totally* known what this is about'. (Official 9) Officials tend to consider the increase in applications for and by foster children (see below) a testimony that attempts at fraud are now finding new channels (Officials 6, 9). Thus, the fact that DNA testing has become a routine procedure may have led to applicants trying to outwit the authorities, as their responses to the 'option' offered by the authorities may be self-selective (see Murdock 2008).

To sum up, the use of DNA analysis has reinforced the rationale of suspicion in immigration administration. Since authorities can use a technique which they expect to provide exact and objective information about family ties to support their decision making, they seem to have become more mistrustful of other forms of evidence and of the applicants. In administrative documents and interviews, authorities make repeated references to false documents and to inconsistent answers given in interviews, as a contrast to the 'accuracy' of information provided by DNA analysis (for example Leitzinger 2010; Official 4, 9). This way of speaking reflects a sort of ubiquity of the suspicion of fraud, something that prevails in immigration policy in many European countries. In such a mood of mistrust and rigid control, division between immigrants with appropriate identity documents and those without becomes increasingly critical.

The Equivocal Role of DNA Analysis

What, then, is the role of information provided by the DNA test in the procedure of family reunification? How significant is it in decision making? Although the Finnish immigration authorities appreciate DNA testing as a provider of accurate and objective knowledge of family ties, the significance of genetic information has remained open to interpretation by the authorities in the Finnish procedure. There has been debate about how much weight should be placed on the result of DNA analysis, as the number of applications for foster children have increased, and the Finnish immigration authorities have adopted a policy that a foster child can receive a residence permit if reliable evidence of de facto guardianship is provided as specified by an amendment to the Aliens Act (SDK 549/2010). In this context, proof from interviews is indispensable. In addition, interviews with the applicants have become generally more pivotal in consideration of family reunification cases with non-existent documents. As an official said, it must be proven that 'there really *is* a family to reunify' (Official 4). More efficient interviewing has been a goal in Finnish immigration policy since the late 1990s, and the development of more thorough interview methods with 'detailed and unexpected' questions was a hallmark of efficiency and equity in 2011 (Maahanmuuttovirasto 2012, 9). The purpose of the interviews is to clarify family ties and provide the 'proofs' of genuine family life – for example in relation to permanence, dependency, sharing of a household and intention to continue family life in Finland. In the eyes of the Finnish immigration authorities, verification of biological ties and proof of 'real' family life are entwined with each other, and in each case the result of the DNA test is considered in relation to the assessment of family life based on interviews, which are seen to provide primary evidence. In Austria, decision making on family reunification cases seems to be based on a similar kind of combination of interview information and DNA analysis, while in Germany the result of a DNA paternity test seems to trump other forms of evidence of family ties (see Chapters 2 and 4). This approach can lead to equivocal and conflicting interpretations of true family ties. In some cases, DNA analysis does not necessarily provide enough support for a favourable decision. This was stated quite bluntly in a press release in 2008: 'A purely biological relationship is not, however, sufficient for a positive decision on a residence permit without the background of a genuine, permanent family life' (Finnish Immigration Service 2008). Hence, positive test results do not necessarily lead to positive decisions. All in all, recent tendencies in the decision making procedure imply that DNA testing is hardly a 'last resort' for the applicants anymore. The 'DNA truth' is becoming increasingly subject to bureaucratic contingencies (see Chapter 4). As the rationale of suspicion has become predominant in Finnish immigration policy, officials are inclined to consider applications for family reunification potentially fraudulent. From this perspective, the 'unexpected' results of the DNA test which may reveal 'family secrets' can also be interpreted as a token of 'deliberately' provided false information, as defined in Section 65(3) of the Aliens Act (SDK 549/2010). This is a significant aspect because such an interpretation by an official can lead to the rejection of the application. In our interviews, immigration officials considered such cases problematic and were reluctant to discuss them. In general, they claimed to be quite 'broad-minded' when evaluating unexpected results. In cases in which the paternity of the alleged father is placed in doubt by the test, the mother may be given a chance to be discreetly heard by the official and to give an explanation of the result.[14] The applicants' informational privacy will be respected, including the right not to know:

---

14 The possibility of revealing 'family secrets' through DNA analysis is acknowledged by the Fact Sheet on the DNA testing procedure. According to the Fact Sheet, 'DNA testing can sometimes reveal information on biological relationships that family members have not been aware of. [In such cases,] the Finnish Immigration Service will reserve an appointment for you and your family members to present your views on the test results'.

> This happens, but less often than I had initially presumed. Mostly they are family secrets the disclosure of which can be avoided, because it is not relevant to know that one of the children was not fathered by the father of the family. OK, it happens. If the father of the family does not know it, it is not our business to tell him. The results of the DNA analysis are formulated so that he cannot read that one of the children is not his. (Official 4)

The safeguarding of revelations was said to be guaranteed because all statements about the tests are sent directly to the Finnish Immigration Service (Geneticists 7, 13). An experienced interpreter (Official 18) told us that he had never seen the test reports, and assumed that the people concerned had not seen them either. However, 'family secrets' can be revealed if the decision is negative and is appealed against in court. In these cases: 'Well, unfortunately it is [revealed] because we have to give the documents to those concerned, and then it is up to the solicitor to decide whether to tell everything … one can hope that the solicitors uses his or her discretion'. (Official 4)

The role of DNA testing is changing in the family reunification procedure, as the recent more restrictive immigration policy establishes new requirements for and obstacles to family migration. For example, Finland and Sweden (like many other European countries) nowadays require that all applicants are present in person in embassies for biometric identification. Usually, they also have to present a passport and a residence permit before the process can begin – even though the passports of the Somali applicants will be rejected. With these requirements, the profile of applicants for family reunification has changed considerably. This is exemplified by the decline in the number of applications submitted for or by family members of recipients of international protection: in 2012, 69 per cent fewer applications of this type were filed than in 2011, and only 30 per cent of the decisions were positive (Maahanmuuttovirasto 2014). Finland was 'the last' European country to cover applicants' travel expenses (Official 4), and since 2012 the journey to Finland has to be paid for by the people themselves. A lawyer (NGO 24) claimed that due to the new conjuncture 'family reunification has become so difficult, even impossible, that people only bring their true family members'.

In Sweden, following a court ruling in January 2012, Somali children without a travel document are treated as exceptional cases and can get a residence permit to join their parents in Sweden (and vice versa) if it can be proven through DNA test that the child and the parent are 'biologically' related (Swedish Migration Board 2012). This use of DNA analysis may imply that not only the family ties but also the identity and credibility of the applicants can be tested and proven by DNA analysis within immigration management. In Finland, similar exemptions to the requirement concerning valid documents were made in 2012 (Finnish Immigration Service n.d.), and while DNA testing is not explicitly mentioned it may in practice acquire this extended role as a means of identification.

Conclusion

Our analysis shows that the Finnish immigration authorities and experts consider that DNA testing provides crucial help for their decision making. There are two reasons for this. First, the technology defines the object of administrative inquiry, family ties, in a clear-cut way on the basis of biological traits of the applicants; it provides a way of acquiring exact, objective knowledge that cannot be manipulated by any individual. It is obvious that the deployment of DNA testing narrows the concept of family to biological ties mediated by DNA, which runs counter to the extension and multiplication of family definitions in legislation and family policy (Heinemann and Lemke 2013; Lippert and Pyykkönen 2012; Murdock 2008). However, due to the features essential to this 'geneticisation', DNA testing can bring the immigration officials 'closer to the truth' in their decision making. For them, the testing is a 'truth machine' in a similar sense as the techniques of forensic medicine in criminal investigations (Lynch, Cole, McNally and Jordan 2008).

However, our analysis shows that the implementation and use of DNA testing are framed by rationales which direct policymaking and administrative practices on family migration. One line of reasoning is related to securing human rights – the right to family, children's rights – and the other one emphasises the need to combat fraud in asylum seeking and family reunification, which reflect a more general trend in immigration policy (Fassin 2005; Huysmans 2000; van der Ploeg 1999). Consequently, knowledge produced by DNA testing machinery can serve two contrasting ends and the use of biotechnology can be considered reasonable and justified from two conflicting but not mutually exclusive angles in the context of immigration policy and administration. In other words, the meaning and impact of DNA evidence is not produced by the test itself but by the political and social context in which DNA testing is used and the results of the DNA analyses are interpreted (see Lynch, Cole, McNally and Jordan 2008).

The emphasis on human rights was a central element of policy reasoning when DNA testing for family reunification was initiated and established in Finland in the late 1990s and early 2000s. Even today, authorities and experts tend to consider DNA test a right, 'the last resort' offered to applicants without proper documents confirming their identity and family ties. In addition, the option of DNA testing was seen to secure the applicants' right to family. From this perspective, the purpose of using molecular biology for verification of family ties is seen as a production of 'truth' that may affirm human rights, support the applicants' claims and secure non-discrimination.

The idea that verification of family ties by DNA analysis may protect human rights of the 'aliens' in the process of family reunification seems to be more explicitly pronounced and perhaps carry more weight in the Finnish policy and administrative rationale than in Austria and Germany (see Chapters 2 and 4).[15] In

15 In Austria, the administrative rationale emphasises that DNA testing is meant to provide evidence to assist in the investigation of the truthfulness of the applicants' claims

practice, officials and policymakers tend to think that such protection is guaranteed by the legal regulation of DNA testing and by a clear administrative procedure in which the testing is completely managed by the immigration authorities (see above).

Another influential framing for DNA testing for family reunification in Finnish immigration policy and administration is a rationale of suspicion, which is built upon a general assumption of untruthfulness on the part of the applicants. A similar frame of reasoning is also evident in Austrian and German policy and administrative rationales (see Chapters 2 and 4), as it seems to be in management of refugees, asylum seekers and immigrants without valid travel documents throughout the western world (for example Bohmer and Shuman 2008; Khosravi 2009).

The idea of fighting fraud was a central motivation and justification for the introduction of DNA testing for family reunification in Finland in the late 1990s, and the rationale of suspicion has become the predominant framework for DNA testing during the past 15 years. This is a general trend throughout Europe. It is clearly indicated by the European Commission Green Paper on family reunification (2011) and by a study on 'marriages of convenience' and false declarations of parenthood by the European Migration Network (2012a). In both documents DNA analysis for family reunification is explicitly put on the agenda of immigration policy, but only in the context of fraud (Heinemann, Naue and Tapaninen 2013). Finland's response to the Green Paper clearly proves that DNA profiling has become an efficient tool in the fight against fraud: 'The preventive effect of DNA testing has for the most part been achieved and no discrepancies are found with the initial information provided' (European Commission 2011). Such a framing is compatible with a general tendency to emphasise containment of 'flows' of refugees and illegal immigrants and to place more restrictions on family migration in European migration policy in the 2000s (Cholewinski 2002; Goenendijk 2006).

Amendments to the Finnish Aliens Act (SDK 301/2004) in 2010 provide testimony of this trend to restrict family migration. The government proposal set the goal for the changes 'to cut down certain pull factors' (HE 240/2009). Six of the nine amendments directly concern family migration. For example, sections 6a and 6b of the law regulate age assessment of alleged minors, and sections 38 specifies that the child has to be underage when the decision is made and not only at the time the application is submitted. It is claimed that the amendments provide legal regulation of already existing practices, but they undeniably place new restrictions on family reunification, specifically on issues related to information and verification. This is quite clear in the Section 36(3) of the law, which states:

> A residence permit by reason of family ties may be refused if there are reasonable grounds for suspecting that the sponsor has received a residence

(Chapter 4), whereas in Finland the purpose of the testing, in principle, is to ensure the applicants' rights. Despite differences in regulating principles, in practice fighting fraud in family migration has become the main objective of DNA testing in both countries.

> permit by circumventing the provisions on entry or residence by providing false information on his or her identity or family ties.

In the service of a restrictive policy, biotechnology like DNA analysis is deployed to control immigrant applicants and may contribute to the suppression of their rights. Because family reunification can be interpreted as a convenient shortcut to Finland for refugee families, they may be suspected of employing deceitful tactics throughout the process.

All in all, DNA analysis for verification of family ties has become a routine of the family reunification process, and the procedure for conducting the testing has largely remained the same for over a decade in Finland. However, the role of DNA testing in family reunification has not remained unaltered. With the changes in immigration policy, the uses and functions of DNA testing have been modified. When the use of a technique of molecular genetics is increasingly framed in the context of prevention of fraud and illicit entry of immigrants, the function of testing is to detect fraudulent applicants, to form a sort of pre-emptive obstacle to hoax applications and even to test the overall credibility of the applicants. As DNA testing serves such functions it strengthens the predominant rationale of Finnish and European immigration policies, both of which emphasise strict border control and containment of immigration 'flows' to Europe and tend to consider an 'alien' attempting to enter Europe as a suspected illegal immigrant (Fassin 2005, 2011; van der Ploeg 1999). In this context, the use of DNA analysis for verification of family ties implies that the bodies of the immigrants and asylum seekers bear the evidence (Aas 2006; Fassin and d'Halluin 2005).

The restrictive turn in Finnish and European immigration policies may also change the weight placed on DNA testing as evidence in the family reunification procedure. There are some indications in decision making by the Finnish authorities that the 'DNA truth' is increasingly being seen as relative to other evidence. The way Finnish immigration officials have tackled the 'phenomenon' of foster children has reinforced a more general tendency to consider information derived from interviews as primary evidence of 'genuine' family life. Consequently, officials weigh evidence of family ties against evidence of family life, and decisions on family reunification are based on the consideration of these two elements. In some cases the results of DNA analysis are 'not enough' to prove the existence of family life, whereas in others the evidence of family life may overcome the negative test results.

The above tendencies in decision making procedure have two implications for the role of DNA testing in family reunification. First, DNA analysis cannot guarantee positive decisions and therefore it is hardly a 'last resort' for the applicants. It seems that the role of the 'DNA truth' is becoming increasingly subject to administrative and juridical interpretation. This leads to the second implication: shifts in the conduct and reasoning of the Finnish authorities indicate that delineations of the 'family' have again become problematic, and the belief that DNA technology would reduce complexity is not as strong as it was 15 years ago. Consequently, even the ontology of the family – What is a family? Who belongs to it? – will remain contested in the politics of family migration.

Chapter 4
Austria: DNA Profiling as a Lie Detector

Kevin Hall and Ursula Naue

Introduction

In this chapter, we discuss the practice of applying DNA profiling in family reunification cases in the context of Austrian immigration policy and management.[1] We analyse how DNA testing is perceived and deployed by different actors in the field (persons subjected to the family reunification procedure, geneticists, state agencies and authorities, international organisations, lawyers, and people working in non-governmental organisations).

The empirical focus of this chapter is on interviews and document analysis. Relevant documents dealing with parental testing in the immigration context such as laws, guidelines, directives, minutes of parliamentary debates, government and expert committee reports, NGO documents (for example websites and press releases) and media reports were reviewed and analysed. The document analysis yielded interesting features of the use of DNA profiling for family reunification, but we consider the 27 interviews we conducted with 30 persons involved as the core data providing access to the actual procedure of family reunification.[2] However, the interviews also shaped our take on the procedure. We were only able to talk to four officials, none of whom were case officers involved in the procedure. We had to deduce the case officers' decision making on the basis of the descriptions provided by NGO interviewees, which we were able to corroborate with the agencies' interviews.

We begin our argument by showing that in practice only refugees are subjected to DNA profiling in family reunification. In the following analysis of the procedure for this group of immigrants, we point out that the Austrian family reunification procedure for refugees is in fact an asylum procedure. In this procedure the family

---

1 Family reunification is not the only application of DNA testing in migration restriction related to Article 8 EHRC. There are many cases within the Dublin II regime where Austria's responsibility in the asylum procedure for families or family members is established with DNA evidence. The majority of these cases are about the family staying together in Austria rather than reuniting the family.

2 Of the 27 interviews 26 were face-to-face (one via phone) interviews and one by email. The number of interviews by actor group was as follows: Federal agencies, 3; regional state agency, 1; geneticists, 6; lawyer, 1; NGOs, 10; reuniting persons, 4; background interviews, 2.

members abroad are the applicants for asylum and the refugee in Austria fulfils an organising role. Our findings suggest that the refugee's trustworthiness emerges as a central problem and the DNA analysis is conceived as a solution. Following the category of trustworthiness into the parliamentary debates (Stenographic Records 2009, 107, 184, 192; see also Motion for a Resolution 2008) on the introduction of DNA testing in Austria, we encountered an atmosphere of mistrust towards refugees. This mistrust is also present in defensive administrative practices of immigration control. Both parties in the procedure, the case officer and the refugee in Austria, turn to DNA profiling for certainty when confronted by conflicting claims about the truth of the refugee's statements on his or her family relations. On the basis of our analysis we suggest that DNA testing becomes a lie detector in the context of Austrian immigration control.

The Subject of DNA Testing

In Austria, there exist two groups of persons eligible for family reunification and thus potentially subjected to DNA testing. The two groups are defined by different laws, rights and procedures.

The Settlement and Residence Act defines the legal category of the 'regular' immigrant and the requirements regarding income, housing and language skills as well as the procedure for reuniting the immigrant with his or her family. The 'regular' immigrant is any non-Austrian citizen who seeks to stay in Austria for more than six months and does not apply for asylum. Article 1 of the Settlement and Residence Act treats asylum as an exceptional case, to which this law is not to be applied. Reunification of families where one family member has been granted a temporary or permanent right of entrance and residence in Austria as a consequence of an asylum procedure is thus treated as an 'exceptional' case.

The procedure for the family of this type of 'exceptional' immigrant, namely refugees, is described in the Asylum Act and handled by the Federal Asylum Agency. This procedure does not lay down any requirements for the family to reunite. Under the terms of the Asylum Act, the family member in Austria becomes the sponsor of the right of entrance and residence of the family members abroad. In this chapter we use the term 'recognised refugee', rather than sponsor, to draw attention to an experience of having to flee from one's home country which is shared by the group of persons under this law. By referring to recognised refugees instead of sponsors we draw attention to the procedural steps refugees have to pass before they are recognised as refugees. In the asylum procedure, the grounds for becoming a refugee are subjected to legal and administrative scrutiny. Only those refugees who comply with the specific legal requirements are recognised as refugees and granted a certain status (for example convention refugees, recipients of secondary protection, immigrants granted residence permits on humanitarian grounds). Each legal status confers different rights and obligations. The term 'recognised refugee' is in this sense simultaneously all-encompassing with respect to legal categories and related to the refugees' experiences.

In our interviews with Austrian immigration agencies we found that DNA testing is practically irrelevant in family reunification cases with 'regular' immigrants. In addition, while Austrian citizens may also want to reunite with their foreign spouses, we never encountered such a case in connection with DNA testing. In fact, this group of family reunification cases never even turned up as a conceivable possibility in our interviews. Our findings suggest that in Austria DNA testing is actually conducted only on recognised refugees who want to reunite with their families.[3]

The Procedure of Family Reunification

Due to the separate legal framings for 'regular' and 'exceptional' immigrants, recognised refugees who want to reunite with their families are treated differently from 'regular' immigrants. As defined by the Settlement and Residence Act, family members of 'regular' immigrants must apply for reunification from abroad. In the ideal case, their documents and the other requirements regarding housing, income, and language skills are checked by the officials in the Residence and Settlement Department, and they are then provided with a travel visa to Austria by the embassy. Once in Austria, they receive their residence permits. DNA testing here turns up as a possibility when the family members abroad have no certificates to prove that they are a family (Article 29 (2)). Possibly due to the socio-economic requirements for family reunification, only families with documents and certificates recognised by the Austrian state file an application under this law.

The Asylum Act defines the procedure of family reunification where recognised refugees are involved. To become eligible for family reunification under this law one first has to be recognised as a refugee. In Austria an application for asylum can only be filed on Austrian territory. Once a person has stated that she or he is seeking asylum that person is brought to one of two primary reception centres (in Traiskirchen and Thalham). The person's fingerprints and a photograph are taken. Case officers interview the asylum seeker for the first time to assess whether Austria is responsible for the asylum procedure under the terms of the Dublin II regime. At this stage the asylum seeker's route to Austria is scrutinised and no questions about the grounds for seeking asylum are asked. These grounds and

3 Neither the Federal Asylum Agency nor the Ministry of the Interior have any official statistics on the number of DNA tests in family reunification cases per year. The forensic laboratories have no reliable statistics either. The Austrian Red Cross (Tracing Services) compiles statistical data on DNA testing in family reunification cases, but as the Red Cross is not the only NGO assisting 'recognised refugees' in their efforts to pass the reunification procedure these data do not represent the total number of such DNA tests per year involving recognised refugees.

further information on the circumstances of the flight and other family members are addressed in a subsequent series of interviews during the actual asylum procedure, as stated by NGO 4.

> Everybody with whom one wants to reunite later on has to be mentioned by the person in the asylum procedure. There was one case where we totally had problems because of this. I don't know why, but this man didn't state this during his own application for asylum, where one has to give general information regarding children, family and everything, name of parents, address in the home village and so on. And for some reason he did not mention one child. This totally caused trouble later on. (NGO 4)[4]

This means that information relevant to the procedure of family reunification is already collected in the context of the asylum procedure before the recognised refugee or the family have even initiated the procedure. The two procedures are fused to form a continuum from asylum to reunification.

During the processing of his or her asylum application the refugee receives assistance from NGOs. After his or her legal recognition as a – for example – convention refugee or as a person needing subsidiary protection, he or she usually turns to the same NGOs again in seeking guidance and assistance for reunification with his or her family.[5] The whole procedure involves the recognised refugee taking an active role in reuniting with his or her family, although he or she is not the actual applicant within the legal procedure. His or her family members abroad are the applicants. In this respect the family reunification of recognised refugees resembles that of 'regular' immigrants as regulated by the Settlement and Residence Act. But at the same time it differs with respect to the roles ascribed to the family members abroad and the immigrant in Austria. Both the Asylum Act and the Settlement and Residence Act outline the procedure only from the point of view of the actions taken by the family members abroad to apply for family reunification. However, the person on whom the family members ground their application is framed in different ways. The term 'sponsor', which is used in the European Commission's Green Paper on the right to family reunification of third-country nationals living in the European Union (2011), is not used in Austrian law. The Settlement and Residence Act instead refers to the 'reuniting person' (Article

4 All quotations from the interviews in this chapter have been translated by the authors.

5 Article 3 of the Asylum Act confers international protection according to the Geneva Convention on Refugees. The family members of recognised refugees with this status can apply for family reunification immediately after the recognised refugee has received this status. Article 8 defines the status of subsidiary protection. The recognised refugee with this status receives only a temporary right of residence and has to apply for a renewal every year. The family members of subsidiary protected persons can apply for family reunification only after the first renewal of this status.

2 (1) 10). This term refers to anyone who is neither an Austrian citizen nor a citizen of the European Economic Area nor a Swiss citizen, who has a residence permit in Austria and from whom this right of residence is extended to another person. Interestingly, this term is absent in the Asylum Act. There is no mention there of either a sponsor or a reuniting person. Referring to the sponsor as the reuniting person the Austrian Settlement and Residence Act implies an active role for the resident immigrant, whereas the Asylum Act makes no mention of the activities of the recognised refugee in the procedure. In our study, we found that the recognised refugee plays an active and central role in the procedure. We therefore extend the term reuniting person, to emphasise the responsibility for the procedure implicitly conferred on him or her by the Asylum Act and taken by the recognised refugee. Our concept of the reuniting person thus takes into account the practical *and* the legal role the recognised refugee has in the procedure.

The activities of the reuniting person comprise tracing relatives in other countries, initiating the process of family reunification, keeping in contact with the Federal Asylum Agency and the embassy involved, and helping to contact and contract the DNA lab. The tracing of the family can take different forms. For instance, a lawyer told us about a case where a mother, after years of fruitless attempts to locate her son, found out about his whereabouts through a chance meeting between her son and friends of hers.[6] In contrast, one reuniting person gave a description of his very structured search for his family abroad:

> For about one and a half years I had no contact with them. And after that I found them. … I did that at a Persian mosque in town. From there I sent a lot of letters. To Afghanistan, to Pakistan, to Tehran, that is Iran. And it took a while but I sent a letter for my parents to every place and I wrote my telephone number on it. (Reuniting person 1)

Reuniting persons turn to the tracing service of the Red Cross or the UNHCR. They also seek the expertise of NGOs concerning the details of the procedure, its requirements, its costs and possible problems. In most cases the NGOs then play an active role in the procedure as they keep in touch with the embassies, the Federal Asylum Agency and the laboratory performing the DNA analysis in the case. They also offer help in organising the family's travel to Austria. Reuniting persons also draw on knowledge and advice from their respective communities.

After having sought assistance from experts for the legal procedure, located the family members, and contacted them, the reuniting person can initiate the actual procedure of family reunification. In a nutshell, the procedure comprises

---

6 In Austria (in contrast, for example, to Germany), lawyers are generally not involved in family reunifications of refugees. This is only the case in cases involving landmark court decisions. Legal counselling and representation in court is provided by NGO counsellors free of charge. In cases where counsellors also act as legal representatives, they mostly have a degree in law.

several steps. To reunite with his or her family the reuniting person has to prove the claimed family relation. Normally, this is done by interviews and legal documents. In cases in which no legal documents are provided, the documents are considered inaccurate by the immigration authorities or the interviews lead to contradictory statements, the DNA test for proving the parental relationship is offered. The reuniting person has to find an accredited lab in Austria. The family members abroad have to travel to one of Austria's embassies and apply for family reunification. The samples required for DNA profiling are taken in the embassy. After verification of the family ties, the family members are allowed to enter Austria. An NGO counsellor described the last steps of family reunification as follows:

> When these people finally arrive here in Austria, well then their first move is to go to a primary reception centre, either in Traiskirchen or in Thalham. There an identification and … a first interview take place. And the folks have to stay there for a few days or they can go directly to the anchor persons. And then they frequently receive a decision within a few months. And then, if the anchor person has asylum they receive asylum, too. Or, if the anchor person has article eight [referring to Article 8 of the Asylum Act which confers subsidiary protection] the outcome varies. So if their own reasons for fleeing their country are stated, and these point in the direction of asylum, then … I recently worked on a case where in the end the child received asylum although the mother had been granted only subsidiary protection. However, most of the time, if the anchor person only has subsidiary protection the others will only get subsidiary protection too. (NGO 4)

Interestingly, the last step of the procedure for family reunification of the family members is similar to the first steps the reuniting person had to take in order to be eligible for family reunification, namely the application for asylum. This application can only be filed in one of the two primary reception centres in Austria. The refugee as a future reuniting person is subjected to an interview, just like his or her family later on after arrival. The Asylum Act frames the family reunification of refugees as an asylum procedure in which the family requests the same protection as is granted to their kin (Article 34). In the following description of the legal framing of the whole procedure of family reunification we identify some issues that agencies, NGOs and reuniting persons have to deal with.

The Legal framing

Officially, 2005 marks the beginning of the deployment of DNA testing in family reunification cases falling under the jurisdiction of the Federal Asylum Agency, although DNA tests were conducted before that year. A representative of an Austrian NGO told us that before 2005 the family members could apply for asylum directly at the embassy by invoking the family procedure. The comprehensive Alien Law amendment in 2005 created a largely new procedure for family reunification in which the embassy only grants an entrance visa in what is known as the embassy procedure. As a result of the amendment, the number of DNA tests in family reunification cases increased considerably. In its search for a local partner for managing and counselling these cases in a pilot project on DNA testing in this procedure, the UNHCR turned to the Red Cross in Austria. In 2006 the Red Cross founded the Competence Centre for Family Reunification, which offered a direct advisory service within regional Red Cross associations. Between 2006 and 2009, the project was funded by the European Refugee Fund (ERF). After the project's funding ended the Red Cross decided to proceed with its counselling service (NGO 1). After this initial testing phase DNA evidence was included as legal option in the Asylum Act in 2009, and this came into effect in 2010 (Article 18 (2); BGBl. I Nr. 122/2009).[7]

From then on, there was a legally regulated possibility of proving the kinship of a family member by conducting a DNA test. The Asylum Act (Article 2 (1) 22) defines family members as parents of underage children, spouses, underage and unmarried children of convention refugees and recipients of secondary protection. Spouses or legal partners (same-sex partnerships and legalised unmarried heterosexual partnerships) of refugees are only eligible for reunification if their partnership or marriage was already in effect in their country of origin. The concept of the family set out in the law is the narrow definition of the nuclear family.

As already noted, the Asylum Act frames family reunification for recognised refugees as an asylum application procedure in which the family requests the same protection the reuniting person receives. The request has to be filed from Austrian territory, which the applicant first has to enter. Thus the application for family reunification begins at an Austrian embassy. The embassy has to grant a visa to a family member of a convention refugee or a recipient of secondary protection for the purpose of applying for the same protection. However, there is a condition, because the embassy has to ask the Federal Asylum Agency for a prognosis of the decision concerning the application for protection (Article 35). The prognosis is given in terms of probability and in no way presents an actual decision on a future application in Austria. In drawing up the prognosis, the Federal Asylum Agency will check if the relationship between the alleged family members and the reuniting person meets the legal criteria for the family tie. In cases where neither the family relationship nor the reuniting person's trustworthiness is sufficiently established, the case officers are obliged to offer a DNA test.[8] In other words, the DNA test is only necessary when the prognosis is negative or inconclusive.

7 If not stated otherwise, we refer to the version of the Asylum Act and Residence and Settlement Act published in BGBl. I Nr. 38/2011. During the course of our research the Residence and Settlement Act, the Asylum Act and the Foreigners Police Act underwent several amendments and changes which, however, did not fundamentally affect what we describe here.

8 The category of trustworthiness is explicitly mentioned in Article 18 (3) of the Asylum Act. We will examine this category later.

The term 'family reunification' does not appear in the Asylum Act at all.[9] However, the possibility of uniting the family is established by granting the family members of recognised refugees a visa under the terms of Article 35, which regulates visa application procedures at the embassies. The family members can then enter Austria and apply for family asylum as regulated by Article 34. In this context, the embassy's request to the Federal Asylum Agency to assess the probable outcome of the procedure entails a pre-procedural checking of the requirements for family asylum. This step is not in itself procedural, and therefore the applicant does not have a legal or administrative possibility of appealing. This limitation is much criticised by NGO counsellors (for example by NGO 5).

Shortly after the amendments to the Alien Law Package in 2009, the Asylum Act was amended again by limiting the possibility of extending the legal status from one family member to the others. Legal status can only be conferred by a family member who has not him- or herself received his or her legal status by way of extension from another family member. Unmarried underage children are the only exception to this rule (Article 34 (6) 2).

> A … relatively new provision … which is controversial to some extent … but which has been introduced to prevent phenomena such as polygamy. Because the problem was that a man can reunite with his wife, and then she can afterwards reunite with her children; but the other way round for example, if a man reunites with his child, then he cannot reunite with his second wife, which means that underage children can reunite any time. But an adult person then cannot derive his/her legal status from a person who also derived his/her status from another person. (Official 1)

Before this amendment, a second wife could derive her legal status from her children who had entered and lived in Austria before her. This is no longer possible, and consequently in such cases children are separated from one parent or the other. At the same time the above-mentioned subsection added a further restriction. In cases where no marriage certificate could be provided before the amendment, the spouse could derive his or her legal status from his or her children. The subsection blocks the reunification of the whole family in these cases. After providing a DNA test to prove the family relationship between the reuniting person, his or her children abroad and his or her spouse, the Federal Asylum Agency requires the certificate of marriage again to establish a direct family relation between the reuniting person and his or her spouse. This is because, according to the Asylum Act, the marriage has to have been in effect before the reuniting person left his or her home country (Article 2 (1) 22). This practice allows for a vicious circle: The reuniting person's documents are rejected. Consequently, a DNA test is conducted

9 The term used in Article 8, 17 and 34 of the Asylum Act is 'Familienverfahren' (family procedure).

to provide evidence missing from the documents, but nevertheless the spouse is not granted a visa without a valid certificate of marriage.

The Embassy procedure

The need to apply for a visa entails that the family has to travel to the nearest Austrian embassy, which is not necessarily located in the country the family lives in. If the embassy is located in a different country the family members first need a visa for that country. The Austrian embassies demand this visa before granting a visa for Austria, and the family members must decide whether they will stay in that country during the process or travel back and forth between the country of residence and the embassy. As the embassy procedure can take up to a year and longer, it is not always possible to prolong the visa for the whole duration. Travelling between the two countries can also be expensive and exhausting, and in the case of travel between Afghanistan and the nearest Austrian embassies in Islamabad (Pakistan) or Tehran (Iran) even dangerous. Either way, once the family has applied for the visa on the grounds of an application for family asylum, the embassy will contact and ask the Federal Asylum Agency to give an opinion on the application.

When the Federal Asylum Agency assesses the probability of the family members' receiving asylum in Austria, the proof of kinship becomes relevant. If the case officer is not convinced of the family relation he or she is obliged to offer a DNA test to the reuniting person and his or her family. In this case the reuniting person seeks a laboratory which can perform the test, and the laboratory contacts the embassy or a fiduciary physician in the country where the family is staying to request assistance in taking the DNA samples. The embassy or the physician then makes an appointment with the family for the collection of the DNA samples, and the family has to travel to that location.

During the pilot project on DNA profiling for family reunification between 2006 and 2009, it was up to the case officers to admit the DNA test as proof at the family's expense or simply reject the whole application because of the lack of a legal obligation to offer the test. Nevertheless, the introduction of DNA tests in the Asylum Act has not removed all ambiguity from the procedure. Moreover, the ambiguity has shifted to the concrete and practical impact of the prognosis, because it is now the embassy that carries out the procedure and not the Federal Asylum Agency.

The embassy will grant or deny the visa on the basis of the prediction and, as explained, the procedure does not give the applicant any opportunity to appeal against the prognosis. Also, sometimes the embassy overrules the Federal Asylum Agency's prognosis. Although the embassy is supposed to grant a visa upon a positive prognosis ('positive Entscheidungsprognose') by the Federal Asylum Agency – which contains the information about the legal status the family members will be granted in Austria – this is not always done. In some cases in which the

Federal Asylum Agency did not demand a DNA test, the embassy has refused to grant the visa and demanded a DNA test itself. An Austrian official told us:

> There are a few cases in which the Federal Asylum Agency, for example, gave a positive prognosis and then the embassies said, well no, we do not believe that, and we want a DNA analysis ... So, such constellations of cases also exist, and yes, then there is a question of how to deal with it. (Official 1)

NGO counsellors told us that in such cases the family is better off satisfying the embassy's demand, because there is no way to appeal against the embassy's decision. The cost of this DNA test, however, is not refunded by the Federal Asylum Agency.

The family members at the embassy are applicants, and their status entails a problem for their legal representation. Due to professional confidentiality codes, the embassy should not pass information to third parties. Any legal representation in the procedure will be based in Austria, and the reuniting person in Austria – who is not considered to be the applicant – will not know about the progress of the procedure. Nor will the embassy pass any information about his or her family members' visa case. The embassy can pass this information to a legal representative authorised by the family members abroad. In this situation communication between lawyers or NGO counsellors and the family members abroad can result in confusion. Giving an adequate authorisation might be complicated due to language differences, illiteracy and absence of means of communication (for example fax, telephone or internet).

Once all relevant samples are in the lab, it will take two to three weeks for the results to be provided to the applicant. After having received the results and the report of the DNA analysis, an official from the Federal Asylum Agency takes a decision on whether the family members abroad get the entry visa or not. The Federal Asylum Agency informs the embassy about the positive prognosis of decision and instructs the embassy to issue the entry visa.

In many cases, the procedure takes several months. After the positive decision, the family members should enter Austria within a period of six months. If this does not happen, the whole procedure has to start again with an evaluation of whether the family members are still eligible for family reunification (for example regarding the age of the children). It usually takes six to twelve months from the filing of the visa application to the family reunion, and if any problems occur in the process the wait can be even longer.

DNA in the laboratories

The Ministry of the Interior and the Federal Asylum Agency both have lists of laboratories considered reliable for conducting DNA testing in the family reunification procedure. The two lists are not identical: the Ministry of the Interior lists four DNA laboratories[10] to be contacted, and the Federal Asylum Agency has three laboratories on its list (Official 1).

During the pilot project on DNA testing for family reunification mentioned above, the Federal Asylum Agency collaborated with the Institute of Legal Medicine at Innsbruck Medical University and the Austrian Red Cross to establish the standards of the test and the requirements for reporting the results of the analysis back to the Federal Asylum Agency. In 2008, the Federal Asylum Agency consulted the laboratory at the Department of Forensic Medicine in Vienna asking for their collaboration. Around that time it also asked a Salzburg lab about their pricing and procedures for DNA testing; however, this lab was not added to the list. Instead, the DNAPRO Institute for DNA Profiling GmbH in Linz was added to the list. In a decree of 2010 the Federal Asylum Agency mentions only the labs in Vienna, Linz and Innsbruck and claims that the expert reports from the other laboratories do not meet quality requirements for reporting. However, the Salzburg lab applies a similar procedure as the three accredited labs regarding most issues addressed here (Official 1, Geneticist 1, Geneticist 2, Geneticist 4, Geneticist 5).

A typical requirement for the forensic laboratory is that it should hold an accreditation according to ISO EN 17025/2005, which certifies quality for forensic procedures. However, it is not mentioned as an explicit requirement in Austria. Nevertheless, all five laboratories conduct DNA testing according to the general prescriptions of the accreditation guidelines, even if they do not hold an accreditation certificate. These guidelines involve a double door system with a code card to enter the lab, a three chamber system for the spatial separation of analysis steps which always has to be passed through in a one way direction, and a strict quality management system (Geneticist 3, Geneticist 4). These criteria reflect the geneticists' preoccupation with (untraceable) contamination. Most Austrian labs are accredited by the German accreditation organisation, since the Austrian Federal Ministry of Economy, Family and Youth as the issuing authority lacks an expert who can check the accreditation criteria. One criterion fulfilled by all the laboratories is that they draw on authorised and court-certified experts either as head of the laboratory or as head of the institute.[11]

Neither the list of three labs provided by the Federal Asylum Agency nor the argument concerning quality requirements of reporting regarding the labs that can be contacted for DNA testing corresponds to the information we obtained about actual practices of contacting DNA labs and conducting the tests. In many cases,

---

10 After consultation with the Federal Criminal Police Office, the Ministry of the Interior generated a list of four laboratories all of which are university departments of forensics: Institute of Legal Medicine, Innsbruck Medical University; Institute of Forensic Medicine, Medical University of Graz; Institute of Legal Medicine, University of Salzburg; Department of Forensic Medicine, Medical University of Vienna (Official 2).

11 The German term is 'Allgemein beeideter und gerichtlich zertifizierter Sachverständiger'. There are at least three other private labs with access to an authorised and certified court expert.

reuniting persons choose freely from the Federal Asylum Agency's list and contact the lab themselves or use an NGO as intermediary and turn up on the site only when they provide the sample. The test pricing and payment modalities differ between the labs. In 2011 the cheapest DNA tests cost between 506 and 550 euros for three persons, with each additional person tested costing 200 euros, and the highest price is 600 euros per person tested. 960 euros for three persons with 300 euros for each additional person is the middle of the price range. All labs have a policy that the applicant has to pay for the test in advance (Geneticist 2). None of the reuniting persons or NGOs we spoke to mentioned comparing the laboratories' prices. They only knew about the one they contracted, which was the laboratory nearest to their residence. As a result there is no competition between the university-based and the private labs conducting DNA testing for family reunification, which reflects a situation somewhere between that of Germany and Finland (see Chapters 2 and 3).

The cost of the DNA test is refunded if the result proves the alleged family relation. Even so, the cost for the laboratory's service represents only one part of the sum required for family reunification. The procedure is very expensive, especially for people granted asylum and for refugees without substantial funds, and the costs are not refunded. Costs for travelling to the embassies, for getting visas for countries with Austrian embassies, for staying in the respective countries/cities, special fees for the embassies' fiduciary physicians taking the samples, and the costs of flights to Austria have to be met. Even though Austrian NGOs collect money for such purposes and even though the NGO 'Asylkoordination Österreich' organises sponsorship and mentoring in the context of the project 'Connecting People', not all families can afford this amount of money. So even if the family is entitled to reunite, this right cannot always be exercised (see Taitz, Weekers and Mosca 2002b).

Once the contract is signed and other requirements like payment are fulfilled, the labs receive a case number from either the Federal Asylum Agency or the Austrian Red Cross. They then try to find a collaborating institution near to the location of the family members to help in the gathering of the samples. During the sample taking the reuniting person in Austria visits the lab only once. The family members abroad have to travel to the location where the sample is taken. Most labs choose Austrian embassies where the embassies' fiduciary physician takes the sample. If sample taking is conducted in an embassy the laboratory sends the kit for sample taking via the embassy's courier service. Afterwards it is sent back the same way. Geneticist 1 describes the standardised procedure as follows:

> In our lab, this is a totally standardized procedure. We take the mouth swab … we send the kits where one can put these mouth swabs really contamination-free. They are air-dried, everything can be labelled properly if the people on-site do it properly, and from this perspective there is a quite high guarantee that the samples arrive here correctly; and then we conduct a DNA analysis in a totally normal way. (Geneticist 1)

Once the DNA is extracted from the bodies, no connection to the identity of the subject remains visible. Thus forensic geneticists insist that this connection has to be established again by other means during the sample taking. For this purpose the Austrian geneticists use a documentation protocol: they often send the tubes for transportation of the sample pre-labelled, one container being used for one person's DNA only. The label then connects the sample to the photos, fingerprints and the passport on the protocol and states a certain putative family role like mother, father, son or daughter (Geneticist 2).

The process of connecting the sample to the person is threatened by the possibility of contamination. Even though the geneticists can easily identify contamination in a sample and figure out when DNA from two or more different persons is present in one sample, the connection to the specific person is hampered by contamination. The connection between the donor and the sample is also jeopardised by the risk of an exchange of samples during the shipping of the samples to the labs. Thus the preferred way of shipping is diplomatic mail from the embassy.

There are many laws in Austria which could regulate use of DNA profiling in the context of immigration, for example the Gene Technology Act (1994/2006), the Data Protection Act (2000/2013), the Criminal Procedure Code (1975/2011), the Security Police Act (1991/2012) and the Medical Professions Act (1998/2010). But they do not, and DNA testing in the context of immigration is not even mentioned in them. However, DNA testing for forensic and diagnostic purposes is regulated, and forensic geneticists draw on these laws for guidance concerning the procedures for protection of the data and samples as well as for their destruction.[12]

From the legal point of view the reports are only to be used in the context of family reunification, and all other data have to be deleted (Official 2). These data must not be accessible from outside the lab (Geneticist 1). The profiles are not to be stored in any Austrian database (Official 2). Other than that, no regulations exist in Austria which directly apply to paternity testing.

Interestingly, the reuniting person plays a different role in the lab than in the overall legal procedure. While the family members abroad are the applicants in the embassy's procedure, the reuniting person is the contractor of the lab in Austria. This means that while the labs are obliged to inform the reuniting person about the results of the DNA analysis the embassies are not allowed to inform the reuniting

12 Labs apply very different policies on sample and data storage. Most laboratories follow general regulations on data storage for medical diagnosis, as laid down in the Medical Professions Act (Article 51 (3) ÄrzteG 1998, latest amendment 2010, published in BGBl. I Nr. 61/2010), and the Criminal Procedure Code (Article 75, 124 StPO 1975, latest amendment 2011, published in BGBl. I Nr. 103/2011). According to the Medical Professions Act information of a diagnosis has to be stored for at least 10 years, while The Criminal Procedure Code rules that the samples have to be stored for at least 30 years; however, some laboratories store them indefinitely so that they can update DNA profiles when new STR systems become available for testing (Official 2; Geneticist 4).

person about the outcome of the visa application of his or her family members, into which the results of the DNA analysis feed as the decisive evidence of the family relation. Furthermore, the reuniting person cannot apply to have the cost of the test refunded. This has to be done by the family members once they have arrived in Austria. The virtual impossibility for the visa applicants of contracting an Austrian lab means that they need help from the reuniting person, so that he or she emerges as a legal hybrid whose active participation is required, whose legal status conditions the outcome of the visa application, but who is in no way entitled to information about the procedure.

Testing Trustworthiness

When the Federal Asylum Agency makes a prediction on a family reunification case, it reviews the recognised refugee's documents and interviews from the asylum procedure. If this review is inconclusive in relation to the prediction, the case officer will offer a DNA test to verify the family status. This offer, however, is made against the background of a possible rejection of the visa application. In addition, an offer of DNA testing may even indicate questioning of the proper status of asylum or subsidiary protection if doubts arise about the recognised refugee's story during the reunification procedure.

At this phase of the procedure the recognised refugee's statements become subject to the case officer's scrutiny. His or her statements from subsequent interviews during the asylum procedure are compared to the statements made in the family reunification procedure. If contradictions and doubts about the recognised refugee's family story arise she or he is offered a DNA test to convince the official of the family status. In this context, DNA testing is seen as enabling a more profound judgement about the recognised refugee's *trustworthiness*.

DNA testing is not automatically offered in cases of missing documents. An official of the Federal Asylum Agency described the decision-making process on whether or not to conduct a DNA test as follows:

> There is absolutely no automatism, it really depends totally on the individual case, and it doesn't mean saying, well, there are no documents and this means that automatically a DNA analysis will be offered; … the general trustworthiness is rather higher; depending on what somebody has said in the first interview; so, if for example a father has already mentioned his five children five years ago in the context of his asylum procedure, and if he has talked about them, maybe also mentioning their names and also the birth dates and so on, and if this story stays the same until the end, we would be more likely to trust him than if he tells us now after five years that he is the father of five children. (Official 1)

The reuniting person has to make references to his or her family consistently throughout all the interviews of his or her asylum procedure. Also, he or she is supposed to know the date of birth of his or her family members. Case officers consider this knowledge essential for a true family relation; not remembering one's wedding anniversary or the birthdays of the children may work against the applicant in Austria. This is different in Somalia, as one reuniting person told us. There is no such thing as a birthday party there, and the date of birth is not considered as important as in Austria (Reuniting person 4). Here DNA tests come into play when the reuniting person's conduct seems perfectly consistent to that person but is confronted with differently valued knowledge about his or her family members. This situation makes the official believe that the reuniting person is withholding information or not telling the truth, and then DNA testing has to convince the official of the family status.

The official has to judge the trustworthiness of the reuniting person on the basis of culturally specific Austrian presumptions, whereby the rationality of the decision is not purely based on conditions outlined by the legal code. In the 2011 version of the Asylum Act the category of trustworthiness (Glaubwürdigkeit) is explicitly mentioned in three articles.[13] Under Article 18 (3), the asylum seeker's trustworthiness is to be evaluated on grounds of his or her cooperation in the asylum procedure. Article 19 (4) states that the asylum seeker must be informed that his or her statements will be used to assess his or her trustworthiness. Finally, Article 60 (2) 2 lays down that the trustworthiness of the asylum seeker's statements is to be established by comparison with documentation on the country of origin.

The Asylum Court clarified further how to understand trustworthiness. In its decision D4 317491-1/2008/10E, trustworthiness is described in performative terms and mostly in relation to the grounds for exclusion. The asylum seeker's statements are not trustworthy if she or he is unable to give detailed descriptions of her or his experience, if he or she contradicts him- or herself, if his or her statements contradict the country reports, or if he or she gives false evidence or on purpose does not give all the information in due time. An Austrian NGO criticised the principle of trustworthiness as a rule of evidence because it was too error prone, due to the circumstances under which asylum seekers are interviewed (Diakonie Österreich 2005, 29–30). In addition, identity papers, marriage certificates and other documentation from certain countries are nearly always treated as false. A member of an NGO explained to us that in countries in which the authorities run short of office materials it is usual to make a copy of someone else's documents, erase the original name on the copy and fill in the new name (NGO 9). Austrian officials regard these documents as frauds. Indeed, one official with working experience in embassies corroborated this procedure. To this interviewee and the embassies these certificates were documents crafted by an official in fulfilment of his duty but with the wrong content. This can have serious consequences for the decision in the asylum procedure. It may also entail that the embassy, despite the positive prognosis, requires a DNA test from the family.

13 Recent versions of the Asylum Act also mention this category, but were not in force at the time the fieldwork was conducted.

Institutionalised mistrust

What is at stake here is not a simple cultural misunderstanding between case officers and foreigners. Rather, practices of mistrust towards recognised refugees are institutionalised. One immigration official expressed this by arguing that the issue of DNA testing in the context of family reunification 'is about the issue of identity theft' (Official 3). The reuniting person is perceived as a potential defrauder rather than somebody who wants to make use of his or her human right of family reunification. This also relates to the practice of case officers who use the DNA test as a technology for revealing contradictions and lies, as a NGO counsellor explained:

> Well, I think that the Federal Asylum Agency is always very much focused on revealing any contradictions … and lies, and on finding any inconsistencies. They are very much focused on that and when tangible issues arise such as diverging dates, names, birth dates, then they become very sceptical and suspicious, yes, and certainly it is helpful when one explicitly states in the first-instance interview what the family situation looks like. (NGO 8)

A current trend among host countries favours more restrictive policies on immigration, especially on family reunification (see John 2003; La Spina 2012). Many countries are imposing stricter requirements on those applying for family reunification, requiring them to provide official documentation and other pieces of evidence to prove their family relation. This is also happening in Austria (see Heinemann, Naue and Tapaninen 2013, 187; Kraler et al. 2013).

While DNA testing was already presented as an option to prove family ties in the 2005 version of the Residence and Settlement Act, DNA testing for family reunification of recognised refugees was only *introduced as a legal option* in the 2009 amendment of the Austrian Alien Law Package. It was passed by the Austrian parliament in October 2009 and came into effect on 1 January 2010. The amendment focussed on follow-up applications, delinquency, territorial restriction and the obligation to register, age assessment and the use of DNA testing. It is noteworthy that there has been no broader public discussion and hardly any media coverage of DNA testing in the context of immigration, although the issue is highly politicised by the right-wing parties (Heinemann, Naue and Tapaninen 2013, 196).

The politicisation of family reunification by associating it with crime, which was originally promoted by the Austrian right-wing parties Austrian Liberal Party (Freiheitliche Partei Österreichs – FPÖ) and Alliance Future Austria (Bündnis Zukunft Österreich – BZÖ), reflects the ubiquitous mistrust in Austria towards immigrants. In the parliamentary debate on the 2009 amendment of the Alien Law Package, the FPÖ and BZÖ consistently linked crime rates to foreigners (Stenographic Records 2009, 107, 184, 192; see also FPÖ 2007). They also argued that some asylum applicants lie and cheat, and that applicants destroy their documents, change their age, their name and their country of origin. In general, asylum seekers and their families are systematically pictured as criminals and defrauders (FPÖ 2007). Maria Fekter of the Austrian People's Party (*Österreichische Volkspartei* – ÖVP), who was then the Minister of the Interior, picked up the idea of DNA testing and argued that DNA profiling should be used in cases in which documents are questionable, fake, or non-existent, or there is no evidence of genuine family ties. She pointed out that refusal of DNA testing results in denying access to the privileges granted to family members of immigrants with legal residence status (Stenographic Records 2009, 184). The amendment was passed without any in-depth discussion of DNA profiling as a means of immigration control.

DNA analysis and immigration control

NGO counsellors perceived DNA testing as embedded in the atmosphere of mistrust described above within the legislature and the accompanying practices of 'defence' exercised by the case officers. This is partly the result of the inconsistent handling of DNA evidence between different case officers. Some officials demand to see the results of the test right from the start, thereby skipping the order of evidence as laid out by Article 18 (2) of the Asylum Act (documentation and interviews first, DNA evidence second). But this mostly works in favour of the families, as one counsellor told us: 'In fact, I do not think that this has led to situations where DNA analyses have shown that the children were not the children of the persons involved; usually these children have been shown to be the children of those persons'. (NGO 5) The same counsellor pointed out a defensive use of the aforementioned Article 34 (6) 2 of the Asylum Act (see above):

> And yes, maybe now they [the Federal Asylum Agency] look for other ways of restricting family reunification, or other arguments. One legal step was the extension to the mother ... for example, if an extension has already taken place, no other extension to an adult person is possible ... formerly this was possible, when they were life partners and if they had a common child it was also possible first to extend asylum to the common child and then from the child to the mother. Since the amendment of the law in 2009, this is no longer possible, which leads to the problem that they [the Federal Asylum Agency] say, so, if you cannot prove that you have been already married, the granting of asylum will be unlikely. (NGO 5)

While the result of the DNA test was supposed to replace official documents in the process of verifying family ties, as argued by the Minister of the Interior in the parliamentary debate (Stenographic Records 2009, 184), and legally fixed in Article 18 (2) of the Asylum Act, it can no longer achieve this for married couples with children whose marriage certificates are not recognised. As mentioned above, in such cases officials of the Federal Asylum Agency demand documentation of family ties but then reject them and require a DNA test as proof of kinship;

however, even a positive result from the DNA test is not enough because the official may then require documentation of the marriage. In this practice, DNA testing is not used to differentiate trustworthy applicants from deceivers but as a 'means of defence' against immigrants. The counsellor quoted above was quite explicit about this:

> Actually, I don't know how often such cases occur, but they occur repeatedly … that, for example, via DNA analyses it might very well be established that the children are common children, but now the marriage is questioned, and then the children are allowed to enter Austria but not the mother … the problem is now, it seems, that for them [the Federal Asylum Agency] too many persons want to enter the country via family reunification, and it is written in the law that only if … a marriage by civil law exists is family reunification possible, and is not possible for unmarried couples. Now they [the Federal Asylum Agency] say that, for example, especially in countries like Afghanistan, if they don't have credible marriage certificates then it is a fact that she is the mother of her child, but it is not documented that a marriage exists. (NGO 5)

This practice has serious implications for the families. The spouses can try to reunite by referring to the Residence and Settlement Act, but they need to meet its requirements, which are practically unachievable by recognised refugees and their families. If the family tries to reunite via the 'regular' procedure the reuniting person has to find a job and meet the income requirements whilst the family members abroad have to learn German. This can mean several years of further separation. Under Article 34 (6) 2 of the Asylum Act children stand a better chance of reuniting with their parents than their parents do of reuniting with one another.

The problem of trustworthiness is not restricted to the legal text but is an extension of a discourse of mistrusting foreigners, especially asylum seekers, and practices of defence against immigrants. Indeed, the debates reflect a presupposition of guilt rather than innocence on the part of the asylum seeker or recognised refugee. Also, their position in society remains precarious as long as they are under the jurisdiction of the Asylum Act. Even in our interviews with NGO counsellors the reuniting person was frequently addressed as an asylum seeker. Counsellors were aware that they were misusing the term, but this shows how the whole process of family reunification for recognised refugees remains within the confines of the asylum procedure. The standard case for the application of DNA tests in the immigration context described here is not an actual reunification. It is an asylum procedure in which the applicants receive asylum.

Truth Machines: DNA Profiling and Lie Detectors

As we argued above, in the family's asylum procedure the trustworthiness of the reuniting person plays a pivotal role in the case officer's decision. The truth of claims about and the truth of the family relation itself are at stake in the DNA testing of reuniting persons and their families, even more so as they are systematically linked with fraud and crime in parliamentary debates about asylum seekers (Stenographic Records 2009, 184; Motion for a Resolution 2008). The use of DNA profiling and DNA evidence in crime investigation and trials has recently been connected to the idea of a truth machine (Lynch, Cole and McNally 2008). The metaphor is not new, since lie detectors were referred to as truth machines before DNA fingerprinting entered forensics. In the following, we juxtapose these two 'truth machines'. We highlight how both truth machines work and confront them with our findings on the use of DNA testing in Austrian immigration policy towards family reunification.

The DNA profile can be considered as a further shift of the level of evidence. It is not a corporeal manifestation of signs of hidden knowledge. Whether a person knows about her identity or not is not an issue in the procedure of DNA testing. Katja Aas (2006) considers DNA profiling a new technology of surveillance of the body by which 'the state and large organizations can establish at least some level of trustworthiness about those marked by suspicion'. Referring to Jonathan Simon, she maintains that these biometric technologies of identification act as 'power without narrative' (Aas 2006, 150). According to Aas, in the process of biometric identification 'profound questions of human nature, character evaluation, danger and trustworthiness are turned into simple, empirical questions of false and positive that can be answered by technology' (Aas 2006, 152). However, our findings suggest that the DNA profiles of reuniting families become entangled in statements about the family and the Austrian requirements for valid documentation, which resonates with what Lynch, Cole and McNally say about the use of DNA evidence in court:

> 'DNA' points the way to guilt or innocence only when it is surrounded by a story containing other claims and evidences. The 'genetic witness' (OTA 1990) speaks for itself only when presented in the form of expert *testimony*, and … interrogation of that 'voice' points to an extended, indefinitely complicated, series of fallible practices through which evidence is collected, transported, analyzed, and quantified. The embeddedness of 'DNA' in stories and sense making has not deterred courtroom participants and legal scholars from making millenarian pronouncements about an ascendant 'truth machine' in a new era of criminal justice. (Lynch, Cole and McNally 2008, 336; emphasis in the original)

Making the body talk

The common feature shared by all kinds of truth machines is that they are supposed to solicit confessions. This is a feature most authors attribute to lie detectors and also to DNA profiling. Lynch, Cole and McNally (2008, 262) say this about DNA typing: 'More often than not its mystique works in favour of the prosecution's efforts to identify suspects, solicit confessions, and pursue convictions'. Kerry

Segrave (2004, 106) affirms this mechanism of truth machines by quoting Berkeley Rice's critique of voice stress analysers: 'Like the polygraph, the voice analyser is also quite effective in eliciting confessions'. To this end, such machines draw on scientific knowledge and claim to make some part of the body, which is invisible to the interrogator, speak the 'truth'.

The body has not always been 'made to talk' in a scientific way. Up until the twelfth century in Europe, judging the guilt of a person was achieved through an act of torture framed as an ordeal in which God would intervene to save the innocent (Segrave 2004, 7–8). In the nineteenth century Cesare Lombroso promoted the idea of *homo criminalis*, the born criminal. Lombroso devised this at a time when the idea of genetic deterioration due to industrialisation and urbanisation was becoming popular. In this conception crime was thought of as deviant behaviour committed by men whose mental state showed itself in their pathological bodies. Criminal anthropology focused on establishing a connection between physical signs on the criminal subjects' bodies and their criminal demeanour. This approach found its limit in connection with criminal women who did not show any physical signs of a criminal nature (Bunn 2012). As a consequence, scientific focus and interest was directed from the outside to the inside of the criminal, and their emotions became the focus of scrutiny. Several instruments were devised to unveil the hidden inner world of pathological emotions, and specific bodily signs were identified that were believed to give away the subjects' emotional states, which were thought of as indicative of a pathological psychology (see Bunn 2012, 177–80). From then on, the signs indicative of a pathological psychology became associated with guilt or innocence.

Among these bodily signs the pulse rate, blood pressure, skin conductivity due to sweating, breathing frequency and others were assessed as ways of detecting deception. All instruments for measuring these signs were invented well before the lie detector was devised. Among these apparatuses were the sphygmomanometer, the galvanometer and the pneumograph, which were synthesised into the black box commonly known nowadays as the lie detector (Bunn 2012, 182). In the 1970s trembles in the voice and in the 1980s brain waves achieved notoriety as ways of unmasking attempts at deception (Segrave 2004, 101–10, 180–82).

Margaret Gibson (2001, 61) emphasises that the lie detector operates on the historical presupposition that truth or signs of hidden knowledge take a corporeal manifestation. The lie detector is thought to turn these signs into a graphically readable form. In this conception the verbal process of making a voluntary or involuntary confession is transformed into a corporeal process. The lie detector does not imagine truth as being concealed within the body, but constitutes the body as if able to speak by translating signs of deception and truth into a graphical form.

Today there exist various methods for conducting a test with the most common lie detector – the polygraph. According to William G. Iacono and David T. Lykken, the different questioning schemes used in the test compare 'the physiological disturbance caused by relevant questions about the crime … with the disturbance caused by 'control' questions relating to possible prior misdeeds' (1997, 427).

Segrave (2004, 185) claims that '[t]he history of the polygraph was the story of a device which never did have any objective evidence to support its claims'. However, lie detectors enact the body as capable of speech when they pose as translators. They transform the language of the body into a graphical sign. The graphemes acquire meaning by establishing distinctions and enabling comparisons (see Rheinberger 1992, 394). The distinction always presupposes a normal state of the body and a deviant state. The deviant state becomes visible only against the background of the supposedly normal state. It is then connected to stress originating in fear of being recognised as a deceiver. Being a deceiver, however involves knowledge of one's deception. In order to lie, a person has to know that what she or he is saying is not true.

DNA testing works quite differently. The test compares the profiles of a trio comprising a putative mother, a putative father and a putative child. The truth of the relation is enacted as a biologic kinship relation in which the profiles 'fit together'. With DNA the body is thought to 'know' its biological ancestry regardless of what the donor of the DNA sample knows of the kin relationships. Aas puts this concisely: 'The whole point behind biometric identification is, in fact, that the mind is deceiving while the body is "truthful"' (2006, 154). From this one would expect that however the results of a DNA test are contrasted with a donor's statements, they can never expose that person as a deceiver. In other words, the test cannot prove the trustworthiness or untrustworthiness of a person. But this is exactly how the test was framed in the parliamentary debates and by the officials we interviewed. In Austria the test does not perform the 'body as password' (Aas 2006, 144–5) as it itself is entangled in 'means of defence', and is always only one piece of evidence alongside others that may establish the trustworthiness of the reuniting person and his or her family. But how, then, is DNA testing used to assess trustworthiness?

There is no scientific evidence that lie detectors actually work (Iacono and Lykken 1997). They have nevertheless been used, in some cases successfully. We argue that their success in eliciting a confession is actually based on the quasi-theatrical performance accompanying them. In addition to what the lie detector is believed to accomplish technically, there is a practical and *performative* side of the truth machine (see for example Bunn 2012; Gibson 2001). An essential part of the socio-technical aspect of the lie detector lies in establishing a dramatic atmosphere surrounding the test. The subject's attention has to be drawn to the machine, and their faith in the objective functioning of the machine and its infallibility has to be evoked (see for example Reid 1947). At the same time the operator of the polygraph has to dissipate the subject's emotional tension when she or he tells the truth (Bunn 2012, 177; Gibson 2001, 69).[14]

---

14 Early in the history of the lie detector, John E. Reid described a technique to accomplish this without recognising that it was the essential part of the functioning of the lie detector: 'the examination is prefaced by a detailed explanation of the importance of the lie-detector test in the case, stressing the fact that if the subject is telling the truth he will

To sum up the characteristics of the lie detector: it is, first, an apparatus to extract a confession; second, this end is achieved in an interaction between at least two persons; third, in such an interaction the interrogator calls upon the power of science to unveil truth and, fourth, makes the interrogated person believe that truth or signs of hidden knowledge take a corporeal manifestation which, fifth, the lie detector is able to make visible due to its scientific powers. So, do these features apply to DNA testing for family reunification?

Our findings suggest that before the DNA test is applied to establish the family relation there has been an interaction between an official of the Federal Asylum Agency and the reuniting person. During this communication the official, for some reason or other, has become suspicious of the reuniting person and his or her identity claims about the persons he or she wants to reunite with or his or her alleged family. This 'ritual for the verification of trustworthiness' (Aas 2006, 150) comprises a component of social communication and interaction. The official suspects that the reuniting person is not telling the truth or is withholding information. To decide whether to trust the reuniting person or not the official offers a DNA test, thus calling upon the power of science. If the test is refused the case is simply closed. There is no need to know whether the reuniting person has lied or not. If the test confirms the parental relationship the reuniting person has told the truth. However, if it yields a negative result his or her claims have been revealed as lies. On the scientific side the DNA test and the lie detector have nothing in common. The former compares non-coding areas of DNA of different persons with one another to establish a biological kinship relation, while the latter compares physiological signs of stress with the normal state of the body. But on the performative side, the results of both partake in the official's scientific enactment of the subject as trustworthy or not.

A DNA Lie Detector in Austria

DNA testing fulfils a dual role in the context of the Austrian family reunification procedure described here. First, testing works as a tool for verification or falsification of information given in the context of the interviews with officials of the Federal Asylum Agency. In this role, DNA testing in practice acts as a lie detector (and can also work as an *arbiter of truth*), even if this is not explicitly addressed as such in Austrian law. Second, DNA testing is mobilised as one 'means of defence' against immigration, especially against refugees.

The DNA test as an *arbiter of truth* is supposed to convince the official of what the reuniting person believes confidently to be true. She or he applies it to underline what she or he knows to be her or his truth as a matter of her or his

willingly cooperate and the instrument will show that he is telling the truth. The subject is also advised that if he is lying the machine will disclose that fact, and then he will be so informed and asked for an explanation'. (Reid 1947, 543) lived experience. A reuniting person expressed this use of the DNA test in the following way:

> Well, she [the official] said at the beginning, well, I have to ... an original certificate and a marriage certificate, so, a certificate of birth and something like that, I have to prepare, and I have done all this. And she said in a straightforward way that this is not enough. Then I said, ok, if this is not enough ... I'll undergo the DNA test. (Reuniting person 1)

In the above statement from the reuniting person another sentiment is expressed, namely that after a long period of interviewing and waiting for a decision he was told that what he had so far presented as evidence had not convinced the case officer. In this context the DNA test was just another necessity he endured, part of the whole procedure in which he invested his hope for a final resolution of the procedure.

This seems to show that the DNA test is used both by the case officers and the reuniting persons as a technology of closure regarding contested knowledge claims. However, case officers draw on it to disclose deception, and reuniting persons use the test to show and substantiate what they think is the truth about their family relations.

With respect to the use of DNA tests as one means of defence, testing fulfils this role in cases where the kinship relation is recognised by the case officers but the marriage certificate is not (see above). First, the *option* of the DNA test to verify the family relation inspires hope of a successful outcome to the procedure. But then, in a next step, this hope is dispelled when the case officer – after positive tests of all the persons involved – demands certificates of marriage again because the recognised refugee's legal status cannot be extended to his or her spouse under Article 34 (6) of the Asylum Act. So, while the children would be allowed to join the recognised refugee in Austria on the basis of the DNA test, their other parent would not (even though the test is positive), confronting the family with the difficult decision of who is to remain separated from the children.

To sum up, in Austria DNA tests in the context of family reunification procedures for recognised refugees work as a kind of lie detector, as an arbiter of truth and as a means of defence. The technique is supposed to help the case officer in deciding whether the reuniting person's statements and claims are trustworthy or not. But obviously, two interpretations of the same step within the not yet initiated family reunification procedure exist. Whereas the representatives of the Federal Asylum Agency focus on fraud, reuniting persons and NGO counsellors highlight the fact that these DNA tests can indeed act as a positive option to establish that the information provided was correct. However, the basis for both practices is not unproblematic, as a DNA test cannot expose lies since this presupposes an infallible knowledge about one's own genetic ancestry. Also, a DNA test cannot corroborate a subject's knowledge about his or her family relations because it enacts the body as knowing, not the subject. Indeed, the subject's knowledge is irrelevant to the test, except insofar as it indicates which other donors should be selected for the purpose of comparing the DNA profile.

In the context of family reunification in Austria, all forms of evidence seem to be implicitly cast in roles which serve to reduce the number of refugees and their families entering Austria – framed by the Austrian Asylum Act and enacted by distrusting and mistrustful administrative practices. Nevertheless, even though the socio-political context for family reunification in Austria is quite defensive, DNA tests contain the positive option of being able to prove a genetically related family, and hence can help persons to reunite after a usually long period of being separated due to war, violence or other dramatic situations. However, not all refugee families are based upon genetic ties (for example families with adopted children or families in the context of same-sex relationships), and the procedure of DNA testing in immigration control may prevent them from reuniting. Also, the procedure – as described in this chapter – is quite exhausting, complicated and expensive and entails many steps which place an additional burden on refugees who have had to flee from their countries of origin.

In addition to the discernible tendency in many countries to reduce the number of refugees accepted, it should not be forgotten that the right to family reunification is enshrined in many international and European (human rights) documents (see John 2003; Taitz, Weekers and Mosca 2002b). It is these documents that should provide the basis for decisions, not only whether frightened refugees give contradictory or erroneous information regarding family members they might want to protect, because the experience of having to leave a country hastily has led to doubts about whether or not to trust Austrian officials. Instead, DNA tests should be used as a positive option and last resort for genetically connected families who cannot prove their legal status via documents because their countries of origin are not perceived as functioning states in a western sense.

Chapter 5
Ethical Aspects of DNA Testing for Family Reunification

Martin G. Weiss

Introduction

DNA testing for family reunification is a highly controversial practice, as it is a battleground of conflicting values and interests which have to be carefully weighed up against each other: on the one hand, the right of the sovereign state to regulate immigration and prevent fraud and child trafficking; on the other hand, the right of the individual person to privacy and family life, widely conceived as a necessary condition for personal autonomy and liberty (see Weiss 2011).

Although it has the potential to speed up the often very lengthy immigration procedure, and may therefore be in the interests of the applicants and even a way to promote the best interests of the children involved, the use of DNA testing for family reunification poses some serious ethical questions. Does the use of DNA tests to prove family ties violate the individual right to privacy of the applicants, especially that of particularly vulnerable underage children? And would other possible ways of investigating family ties, most obviously extensive interviews by immigration officers, be less harmful? Are DNA tests for family reunification a form of indiscriminate criminalisation of immigrants, at least when their data is automatically stored in forensic databanks? What about the psychological distress caused to those who may discover that they are not biologically related to their parents or children? What about the right not to know?

The answers to these questions depend not only on the kind of samples and data collected and on how they are handled and stored, but also on what is understood by privacy and informed consent, something that is central to privacy issues in the medical context and also in the non-medical context of DNA testing for family reunification.

The ethical problems posed by the use of DNA analysis for family reunification are complicated further by the fact that the agents involved in the procedure are not only the state and the individual immigrants, but also a third entity: the family, which moreover is granted special protection in many international treaties and national laws. What idea of family underlies the practice of DNA analysis in the context of family reunification? And is this technique establishing a sort of dual standard in the way in which society views families – on the one hand the biological ties of immigrants, and on the other social relations between native citizens?

In addition to these questions concerning the protection of privacy and family life, DNA testing for family reunification raises the broader issue of biologisation of social relations, as this practice at first glance seems to reduce the family to a merely biological notion. Does this practice produce the immigrant as a form of 'bare life' (Agamben 1998), that is to say a rightless object of state sovereignty, or must it rather be seen as a symptom of an ongoing general, perhaps even empowering, biologisation of the social that influences family reunification procedures among other spheres?

Privacy Issues: DNA Tests Versus Interviews

'The scholarly literature prescribing ideal definitions of privacy is extensive and inconclusive'. (Allen 1997, 34) Although Anita L. Allen is probably right, we cannot forego the notion of 'privacy', which has become one of the most influential concepts not only in contemporary ethics but also in legal codes, and is today one of the main ethical issues connected with the use of DNA analysis for family reunification. This is so not only because the tissue samples taken to perform the kinship test can, if not destroyed, potentially be misused to gather all kind of information about the genetic asset of the applicant (including dispositions to certain diseases and phonotypical features), but also because the information about a person's biological relatedness is, in and of itself, very personal.

The right to privacy is closely related to the notion of liberty or autonomy, the core value of modern liberal societies, because privacy is widely perceived as a necessary condition for personal autonomy (Mindle 1989). For the philosophical doctrine of liberalism, the private sphere with the individual at its centre is therefore a necessary condition for personal freedom and autonomy. All public or state interference in this sphere can be justified only where the exercise of one's liberty would violate the liberty of another. In John Stuart Mill's (2003, 13) famous phrasing:

> The only purpose for which power can be rightfully exercised over any member of a civilized community, against his will, is to prevent harm to others. ... The only part of the conduct of anyone, for which he is amenable to society, is that which concerns others. In the part which merely concerns him, his independence is, of right, absolute.

That autonomy (understood as self-government) needs privacy means that self-government must take place undisturbed by the outside world. Freedom necessitates isolation from outside interference; in other words, freedom needs privacy. Thomas M. Cooley consequently defined the right to privacy as the 'right to be let alone' (Cooley 1880, 29). In their famous essay 'The Right to Privacy' Louis D. Brandeis and Samuel Warren (1890, 196) state, with reference to Cooley:

> The intensity and complexity of life, attendant upon advancing civilization, have rendered necessary some retreat from the world, and man, under the refining influence of culture, has become more sensitive to publicity, so that solitude and privacy have become more essential to the individual; but modern enterprise and invention have, through invasions upon his privacy, subjected him to mental pain and distress, far greater than could be inflicted by mere bodily injury.

As a condition of liberty, privacy is therefore an established and widely acknowledged moral right, although debate continues about what exactly privacy means. While authors like Edward J. Bloustein, located in the classical liberal tradition, equate privacy and human dignity (Bloustein 1964), Charles Fried and W.A. Parent agree that privacy is a matter of control over personal information, that is to say *the right of the individual to decide which information about herself may be made available to others and under what circumstances*. However, Fried and Parent differ in the rationale they give for the importance of privacy. For Fried, privacy is the condition for a personal relationship, as he sees the condition of intimacy in control over personal information, meaning the possibility of selective disclosure: 'to respect, love, trust, feel affection for others, and to regard ourselves as the objects of love, trust, and affection is at the heart of our notion of ourselves as persons among persons, and privacy is the necessary atmosphere for these attitudes and actions' (Fried 1970, 477). For Parent the importance of privacy, understood as a 'condition of not having undocumented personal knowledge about one possessed by others' (Parent 1983, 269) is instead based on three arguments:

> First of all, if others manage to obtain sensitive personal knowledge about us, they will by that very fact acquire power over us. Their power could then be used to our disadvantage. … Second, as long as we live in a society where individuals are generally intolerant of life styles, habits, and ways of thinking which differ significantly from their own, and where human foibles tend to become the object of scorn and ridicule, our desire for privacy will continue unabated. … Third, we desire privacy out of a sincere conviction that there are certain facts about us which other people, particularly strangers and casual acquaintances, are not entitled to know. … Those of us educated under … liberal ideology feel that our lives are our own business (hence the importance of personal liberty) and that personal facts about our lives are for the most part ours alone to know. (Parent 1983, 276)

The ethical problems of interviews

Before addressing the privacy issues posed by DNA tests for family reunification, I want to look briefly at the privacy issues posed by the well-established alternative to prove family ties, extensive interviews, which some authors consider ethically preferable to DNA tests (Murdock 2008). In fact, in family reunification procedures sensitive personal information comes under intense scrutiny not primarily in the context of DNA testing, but much more directly than a DNA test would entail (especially if the data is collected with parsimony and the samples promptly destroyed) in the context of well-established immigration interviews aimed at proving the existence of familial bonds. Although the questions posed by immigration officials to discover a marriage of convenience (Pöry 2010) or child trafficking (Crawley 2012) are, for obvious reasons, not officially available, it is clear that to prove the existence of intimate private relations – and this is what family life is about – the interviewing officer has, by definition, to violate the right to privacy of the applicants. If one adopts Parent's definition of privacy quoted above, stating that the right to privacy consists in the right to have control over one's personal information, that is in *the right of the individual to decide which information about herself may be made available to others and under what circumstances*, it is clear that in family reunification procedures the applicants are coerced to make personal information about themselves available to others, as to refuse to 'cooperate' with the immigration authority would lead inevitably to a rejection of their application. In addition to this general violation of privacy due to the fact that the applicants have no real choice about disclosing personal information if asked by the immigration authorities (as they have no real choice about whether to take the DNA test), immigration interviews also violate informational privacy because of the type of information they require applicants to disclose. Not only the fact that the applicants are forced to speak about their intimate relationships but also the type of information they are required to give is an intrusion into their privacy. It is easy to see from the guidelines accompanying EU Directive 2003/86/EC on the right to family reunification the kind of questions that are likely to be asked. Sham marriages can be revealed by interviews in which 'the couple are inconsistent about their respective personal details, about the circumstances of their first meeting, or about other important personal information concerning them' (Pöry 2010, 24). A further issue concerning privacy in the context of immigration interviews is that the personal information collected and documented is not treated as confidential, as the immigration authority commonly 'reserves the right to disclose information to other government departments or agencies, local authorities, international organisations and other bodies to enable them to carry out their functions' (Crawley 2012, 52).

With regard to informational privacy, DNA tests to verify family ties are therefore probably less intrusive, and hence less problematic, than scrutinising interviews, at least if the tissue samples and the genetic data collected are used only for the specific family reunification procedure as is the case in Finland (see Chapter 3).

This means that, in regard to the old technique of interviews on the one hand and DNA testing on the other, it has to be emphasised that from an ethical standpoint the older technique is probably not less problematic than DNA tests because its potential to harm personal privacy is equally great, especially when intensive interrogation techniques are used on underage children. Thus interviews are not automatically the ethically less challenging technique, as Terra Rica Murdock (2008, 1528) seems to think when she claims:

> Children as young as three could be interviewed about specific memories with the interviewer helping to narrate their experience without influencing their disclosure. Once children reach the age of five, they are generally better able to recall kinship outside the context of nuclear family (aunts, first cousins, etc.). An interviewer may be able to discern whether a child actually knows the person to whom they are claiming to be related.

Although it may be less expensive to interview children than to carry out DNA tests, as Murdock claims (and even this calculation is questionable given the ever decreasing costs of DNA analysis and the need to pay specially trained interpreters), from an ethical point of view interrogations of underage children can hardly be advocated. This is because interviews of this kind put the children under immense psychological pressure and may lead to worse harm than DNA testing. If an unexpected DNA test result ('true' or 'false') is able to produce severe harm to family relations, a negative decision by the immigration service based on the (maybe misinterpreted) interrogation of a child, resulting in the denial of family reunification, creates an unbearable burden on a child who may now for the rest of his life be responsible for the separation of his parents. A false, that is to say misinterpreted, interview with a child is therefore ethically probably more problematic than a false DNA test. So contrary to the opinion of Murdock, from an ethical point of view it has to be stressed that interviews (at least with small children) are more problematic than DNA tests in family reunification procedures due to the enormous stress and psychological pressure they represent:

> In relation to children and young people who are not subject to immigration control, but who come into contact with professionals such as the police, the courts, legal representatives and social services, it is widely recognised that it is important to minimise the number of interviews and assessments to which they are subjected. This is acknowledged to be in the best interests of the child. (Crawley 2012, 42)

And this is especially true in a situation in which the future of the family may depend on the answers of the child.

Furthermore, a secondary aim of the family reunification process to which Murdock refers, the discovery of victims of child trafficking, can probably be achieved more easily and with much less distress for the child by DNA testing than by intensive questioning of children who may have been intimidated by their traffickers.

In the Convention on the Rights of the Child (CRC*)*, signed by all member states of the United Nations except the United States and Somalia, as well as in the ethical literature on children's rights, there is broad consensus that the leading principle when dealing with children must be the 'best interests of the child' and that as far as possible even very young children must be included in every decision-making process that affects them directly (CRC Article 12; Archard 2011; Kopelmann 2004).

The discussion about interviewing children in immigration procedures therefore reflects a certain tension between the principle of the best interests of the child, which seems to demand avoiding distressing interviews, and the principle of participation, which seems to favour interviews. Although interviewing children is ethically problematic, 'the right of children to be heard is a fundamental principle in international law. The Convention on the Rights of the Child provides that a child who is capable of forming their own views has the right to express those views clearly in all matters affecting them' (Crawley 2012, 45).

In order to evaluate the pros and cons of DNA testing versus interviews to prove family ties in family reunification procedures, one therefore has first to find out what is in the best interests of the child.

As the CRC, in accordance with the General Declaration on Human Rights, explicitly stresses the right (of the child) to (his) family, family reunification is certainly in the best interest of the child as long as there is no danger of domestic violence.[1] Furthermore, in the context of immigration 'the best interest principle means that children and young people who are subject to immigration control need a timely resolution to their case and certainty about the future' (Crawley 2010, 34).

As far as this 'no delay' principle is concerned, and as a less stressful and less intrusive procedure, DNA testing is probably preferable to extensive and distressing interviews about private life. However, DNA testing for family reunification poses ethical challenges of a different kind.

The ethical problems of DNA tests

DNA tests, unlike interviews, cannot prove the existence of a social family, that is to say what is commonly meant by the word 'family', but (only) biological kinship. Thus the decision to use DNA tests or interviews to prove alleged family affiliations depends first of all on the type of family a state wants to ascertain. If the focus is on a social notion of family, as is the case in Finland (see Chapter 3), interviews will be unavoidable. Germany instead has placed the focus on biological ties, probably more for practical reasons related to the large numbers of applicants and the principle of no delay than because of theoretical considerations (see Chapter 2).

But if one has decided that what should be proven are biological ties and not social ones, what is the problem with collecting genetic samples and data from immigrants in family reunification procedures? The answer to this question depends mostly on how the collected tissue samples and data are handled and stored. In the

1 'As a guiding concept, reunification with parents or other carers will generally be in the best interests of the child or young person'. (Crawley 2012, 133)

context of DNA testing for family reunification it is therefore extremely important that the data and tissue samples collected are destroyed immediately after the test, as is common practice in Finland (see Chapter 3). If it is not clear what happens to the collected tissue, as in Austria (see Chapter 4), or where the data can be stored in forensic databanks (see Chapter 2), difficult questions of genetic privacy arise. Allen distinguishes 'informational, decisional, physical, and proprietary dimensions' (Allen 1997, 31) of genetic privacy. According to Allen, genetic privacy is normally associated only with informational privacy in the sense of 'the claim of an individual to determine what information about himself or herself should be known to others' (Westin 1994, 54) based on the notion of individual autonomy or self-determination. However:

> The genetic privacy concerns heard today range far beyond informational privacy to concerns about physical, decisional, and proprietary privacy. Briefly issues of physical privacy underlie concerns about genetic testing, screening, or treatment without voluntary and informed consent. In the absence of consent, these practices constitute unwanted physical contact, compromising interests in bodily integrity and security. Decisional privacy concerns are heard in calls for autonomous decision making by individuals, couples, or families who use genetic services. A degree of choice with regard to genetic counseling, testing, and abortion are a requirement of respect for decisional privacy. The fourth category of privacy concerns proprietary privacy, encompasses issues relating to the appropriation of individuals' possessory and economic interest in their genes and other putative bodily repositories of personality. (Allen 1997, 34)

In the context of DNA analysis for family reunification, the issue of genetic privacy arises in connection with the procedures regulating the storage and possible forensic use of tissue samples. Whereas the data collected to identify familial relations are relatively unproblematic, as they usually do not contain information about the phenotype or disease dispositions, this highly sensitive data can be retrieved at any time from the stored tissue. In this regard the tissue samples taken for DNA testing for family reunification should be destroyed after the collection of the data necessary to establish the familial relationship in dispute, as is the case in Finland (see Chapter 3), and not stored for up to 30 years and possibly used for criminal investigations as the German system allows (see Chapter 2). Inconsistently, in the non-medical, forensic context of DNA testing genetic privacy is given much less attention than in the context of medical intervention, where strict ethical frameworks and professional ethics are in place to prevent the misuse of data. Ideally, in the medical context the storage of genetic material or genetic information is allowed only with respect to the fundamental principle of informed consent. This consent normally has to be given in writing, and serves to ensure that the person concerned is informed about the exact amount, nature and scope of the collection and processing of her data. The forms on which the informed consent is recorded have to specify that the consent is voluntary and can be retracted at any time, and must explain what (legal) consequences a refusal to give consent may imply. Informed consent given under coercion is considered invalid (Beauchamp and Childress 1994; President's Commission for the Study of Ethical Problems in Medicine and Biomedical and Behavioural Research 1982; Wear et al. 1997).

In the context of DNA testing for family reunification these standards can hardly be met, for several reasons. The large body of literature on informed consent in the medical sphere stresses that the way in which the patient is advised in order to be able to give truly informed consent should not assume an idealised, reasonable person, but must rather take into account the specific situation of the actual person whose consent is desired. Thus, it is ethically not sufficient to provide the related information about the meaning and possible consequences and alternatives in a way that an average person (or even worse, an average physician) would understand, but in such a way that the person involved is able to understand what she is consenting to. It is probable that this sort of ethically sound information on the meaning, the possible alternatives and the consequences of DNA testing in establishing a genetic relationship may be especially difficult in the context of family reunification cases where the persons involved come from different cultural backgrounds and may often have very little previous knowledge of genetics.

A telling example that illustrates the difficulties involved in getting informed consent for DNA analysis in family reunification cases is the legal regulation of this matter in Germany (see Chapter 2). The focus of the relevant law – the Gendiagnostikgesetz (Act of Genetic Diagnostics) – is on the right to informational self-determination, with the aim of protecting individuals from abuse of their genetic information in the medical context. But when it comes to DNA analysis for family reunification the Act explicitly denies the applicants the right to retract their consent, the right to have their data and samples destroyed, the right not to have their data used for secondary purposes (mainly forensics), the right to be informed about the possible health risks (including the possible psychological shock of learning that supposed relatives are not biologically related), and finally the right not to know (Gendiagnostikgesetz Section 17, Abstract 8).

By explicitly denying the applicants (who in most cases are not German citizens) these fundamental rights the law potentially establishes a dual standard, as (mostly foreign) applicants are stripped of rights granted to German citizens. Furthermore, the applicants are de facto indiscriminately criminalised by the storage of their data and tissue samples for forensic purposes at the discretion of the authorities. It is noticeable that all the rights denied to applicants in Germany are explicitly granted to applicants in Finland (see Chapter 3).

In regard to the German denial of a right not to know (Andorno 2004; Chadwick et al. 1997), to whom one is biologically related, it is also alarming that possible health risks of DNA analysis in the context of family reunification, which mainly means the possible psychological shock arising from learning that a biological relationship one had never doubted does not exist, are not addressed in the German legal framework of 2010 and explicitly denied in the ethical guidelines published a year later. According to these guidelines, there is no need for information on any health risk because there is no such risk (see Gendiagnostikkommission 2011).

An especially delicate aspect in this context is the problem posed by the treatment of underage children. These children play a key role in family reunification cases, as their biological relatedness to both alleged parents is their ticket to being reunited. In the specific case of DNA analysis for family reunification, it is not possible to bypass the need for informed consent of the children by referring to the consent of the parents, on the simple logical grounds that before the test, the alleged parents cannot be considered the legal parents of the child; the whole purpose of the DNA analysis is to establish that they are in fact the parents of the child. On logical grounds, one cannot ask parents whose parenthood is in doubt to give their consent to a DNA test on the child in question, in order to prove that they are in fact the parents and are entitled to decide for their child. Nevertheless, this is exactly what the German and the Austrian authorities do when they allow the alleged parents to give informed consent to DNA testing of their alleged children, in order to prove their parenthood (see Chapter 2 and 4). In this practice, at least in Germany and Austria, a more social definition of the family (the people showing up to ask for reunification) is seemingly used as a way to get the legally necessary informed consent to demonstrate the existence of 'true' biological family relations, which then trump the formerly assumed social relation. This may be an example of how the abstract products of science and technology are ousting our natural experience (Husserl 1970), on which they nevertheless have to rely in a dialectic movement of exclusion by inclusion, as to negate the existence of a (biological) family (to deny immigration) it is necessary to affirm the existence of a social family (to get informed consent to the DNA analysis) first. This means that the ethical standards of informed consent can hardly be met in family reunification procedures. Nevertheless, I would argue that balancing the ethical difficulties posed by DNA testing against the possible harm caused by intensive interviews, DNA tests are less problematic because they are less distressing and less intrusive than interviews, as long as the consent to the test is as informed as possible (meeting the medical standards for informed consent) and the data and tissue samples collected are used only for the specific reunification procedure and destroyed immediately after its conclusion. If these standards are met, DNA testing, which accelerates the reunification process, may even be considered a better means to comply with the Convention on the Rights of the Child, which in Article 10 (1) explicitly states that family reunification cases involving children 'shall be dealt with by States Parties in a positive, humane and expeditious manner'.

As already mentioned, the Convention also emphasises the right to family life. It is this right to family life on which the right to family reunification and the means used to ensure this (whether interviews or DNA tests) are based: 'The right to family reunification flows from the need to protect the family as the natural and fundamental unit of society, as recognised by the Universal Declaration of Human Rights and the International Covenants of 1966 on Civil and Political Rights and on Economic and Social Rights' (EUMC 2000, 96).

But what exactly is the object of family reunification procedures? What kind of relation between persons must exist for them to be allowed to reunite? What is a family? And why is it institutionalised and granted special protection? Because it provides the necessary private sphere individuals need to develop and practice autonomy and liberty? Or because it is the 'natural and fundamental unit of society'?

The Right to Family Life

'There is no standard definition of family' (UNHCR 1999, 102), and there never was. The Greek term *oikos*, translated into Latin as *familia*, originally designated the household, composed of the members of the extended family, slaves and domestics and ruled by the *pater familias*, who was entitled to absolute sovereign power. Although this account is valid only for the western notion of the family and only captures its main features, the traditional narrative (on which much of modern family policy is still implicitly based) reveals that this did not change much until the nineteenth century, when a new notion of the family began to emerge in Europe. The modern family started to develop at the beginning of industrialisation, with the division of labour and the related process of urbanisation. Individual men began to leave their extended families in the rural parts of Europe and to migrate to the cities in search of work. In the cities the organisational form of labour became an integral part of industrialisation, leading to a separation of working space from living space. This spatial separation of the sphere of work from the sphere of life, the separation of production from reproduction, came with a differentiation of gender roles that restricted the man to the role of bread-earner and the woman to that of mother and homemaker. The idea of the core family, an intimate, private relationship between a father, a mother and a child, began to become the main notion of the family and was protected by the right to privacy that developed simultaneously. Based on this situation, in the nineteenth century the idea emerged that the family was a phenomenon of natural law, preceding all legality, and that in its function of reproduction and socialisation of autonomous and responsible citizens the family constituted the fundamental unit of society. This notion quickly became a normative ideology, and it continues to permeate national laws and international declarations today. Thus in 1994, the year of the family, the General Assembly of the United Nations declared the family to be 'the foundation of human society and the source of human life' (UNHCR 1999, 103). This idea is also alive in the statement of the German philosopher and former German constitutional court judge, E.W. Böckenförde, that state and society rely on foundations that they cannot themselves produce (Böckenförde 1976, 60). Accordingly, the family was perceived as inviolable and deserving of legal protection in a similar way to the natural freedom/privacy of the individual. The right to family life is today unquestioned. Both the General Assembly of the United Nations and the catechism of the Catholic Church treat the family as the fundamental element of human society, and the special protection of the family is an integral part of most modern legislations. Interestingly, this ideology was not only propagated by liberal societies but also in socialist states. Thus, Article 38 of the constitution of the former German Democratic Republic guaranteed special protection to marriage and family life.

The key role accorded to the family in modern, industrialised, highly complex societies in which economic production takes place outside the family is grounded in the notion that the societal subsystem of the family fulfils the essential functions of reproduction and socialisation of the members of society. In this (admittedly conservative but still influential) view, the family accomplishes the quantitative and qualitative production of human capital (Berger and Berger 1983). Although family life in liberal democracies is seen as part of the private sphere, the cumulative outcome of private decisions concerning family life is of public interest due to the fundamental social and productive functions that are attributed to the institution of the family. In the context of neo-liberalism family politics are therefore an increasingly important part of modern politics or biopolitics. Although in most western societies less traditional, that is to say less instrumental, concepts of the family (including homosexual partnerships and patchwork families) have emerged, the general rules of family reunification tend to reinforce the traditional instrumental perspective on the family by continuing to treat its paramount function as the production of human capital. Thus the UNHCR has (probably meaning well, but implicitly reducing the immigrant and the family to a mere function of the country of arrival) pointed out that facilitating family reunification is not only in the best interest of the immigrants but also in the best interest of the accepting states:

> Experience has shown that the family unit has a better chance of successfully reintegrating in their homeland or integrating in a new country, than do individual refugees. In this respect, protection of the family is not only in the best interest of the refugees themselves, but is also in the best interest of states. (UNHCR 1999, 118)

What makes DNA testing for family reunification even more problematic in this context is the fact that in modern liberal societies, neither the features nor the functions of the family (allegedly needed by society) are bound to the existence of biological ties between family members. In most western societies the alleged functions of the family that are seen as indispensable for society, procreation and primary socialisation, can also be carried out by same sex partnerships or patchwork families whose members are not biologically related. Although this is also increasingly being acknowledged in matters of immigration (Holland 2008) – as we can see in the Finnish example, where DNA tests are not the crucial element (see Chapter 3) – immigrant families are often still reduced to their biology, as for example in Germany, where DNA tests are being used in an increasingly routine way (see Chapter 4). This requirement of a biological link between family members cannot be justified by the preservation of the alleged vital functions of the family and is diametrically opposed to family recognition policies in most host countries, which emphasise social rather than genetic ties. This results in the ethical problem of establishing a 'dual standard of family recognition' for native citizens and immigrants (Murdock 2008). Socially, this focus on genetic links also means that alternative forms of family are devalued and seen as secondary; they are not recognised as 'true' family ties. The practice therefore fails to consider not only the different definitions of family of the immigrant/refugee culture, but also the definition of family held by liberal democracies. So whereas the western societies seem to embrace more and more liberal definitions of the family and are abandoning the biological concept, DNA tests for family reunification tend to work against this general notion. In cases where a positive DNA kinship test is the only way to family reunification, it is the common social definition of the family in western liberal societies that is denied to immigrants, as the majority of western liberal societies share a pluralistic and mostly socially defined notion that includes patchwork families, adopted children, and same sex families, whereas immigrants in these cases are de facto reduced to their biology.

On the one hand, this biologisation of the family seems to strengthen its traditional social function (procreation) and patriarchal structure, as (married) partners can prove that they are 'truly' related only through their biological offspring, which thus becomes the 'material proof of their conjugal love'. This idea echoes and performatively naturalises (Butler 2006) the traditional Christian definition of marriage and silently reintroduces the concept of 'illegitimate children'. Western civil societies have fought for decades to abolish the stigmatisation of 'illegitimate' children conceived outside official marriage. The return of the distinction between legitimate and illegitimate children, implicit in the current practice of immigrant family reunification, is at the very least anachronistic and is in stark contrast to the views of the general public in western societies. In an attempt to mitigate this anachronistic effect of DNA testing for family reunification, which is (un) intentionally fulfilling the function of a test of marital fidelity, thus posing serious gender problems, in France the law does 'not test paternity in order to avoid embarrassing revelations about fidelity in past relationships' (Murdock 2008, 1511), but only the biological kinship between mother and child. Thus by reducing the immigrant family to biological ties, the practice of DNA testing for family reunification tends to promote and stabilise the patriarchal nuclear family, 'namely a legally married husband and wife and their dependent, unmarried minor children' (Taitz, Weekers and Mosca 2002b, 24) and the connected traditional gender roles criticised not only by feminists but also by most of the western countries performing DNA testing for family reunification when they are faced with the situation of women in many of the immigrant cultures.

On the other hand, the biologisation of the family resulting from this practice could be seen as an integral part of a relatively new empowering tendency within a general biologisation of human relations that is also visible in the emergence of consumer genomics (Prainsack et al. 2008) and forms of 'biological citizenship' (Petryna 2002) or 'genetic citizenship' (Kerr 2003; see also Chapter 6).

Conclusions: Autonomy, Privacy, Family Life and the Empowering Aspects of DNA Testing

DNA analysis for family reunification represents a crossover of highly contested and widely discussed ethical and political issues, ranging from the concepts of family and privacy to immigration and biologisation.

Although there is an ongoing debate about whether states have a moral right to limit immigration into their territories at all (Walzer 1984; Wellmann 2008; Lister 2010), to control immigration is widely considered a legitimate right of the sovereign state. This is expressed, for instance, in Nicolas Sarkozy's explicit claim 'that selective immigration was an "expression of France's sovereignty"' (Murdock 2008, 1508). Operating in a sphere remote from philosophical discussions, immigration control is an unquestioned practice in the realm of *realpolitik*, codified in a number of international conventions and treaties that often explicitly address the evident contradiction between immigration control and the right to privacy and family life. Thus Article 8, Paragraph 1 of the European Convention on Human Rights establishes the right to private and family life, only for this same right to be restricted in paragraph 2 in the name of public security or economic well-being.

In this context, Murdock rightly suggests that national legislation on family reunification should at least match the standards set by the Convention representing the basis of the ruling of the ECHR, which on her account would probably overturn any decisions of national immigration agencies which rely only on biological ties and fail to take into consideration social parenthood (Murdock 2008, 1519); what is protected by the right to family life codified in the convention 'is the actual living together of … parents with their children, also those from before the marriage who are not related to one parent, when they are taken into the family' (Opsahl 1973, 188).

As we have seen, Murdock therefore suggests replacing DNA tests in family reunification procedures with interviews with the applicants, including small children. Although interviews are probably the best way to accomplish the requirements of the Convention, that is to prove that there are real (social) family ties, because of the violation of privacy they constitute and the distress they cause interviews are ethically more problematic than DNA tests, provided that the genetic data collected is used only for the specific reunification procedure and the tissue samples are promptly destroyed. While DNA tests thus are preferable to interviews from an ethical point of view, they have the epistemological disadvantage that they only reveal biological kinship, whereas true social family life (which could be that of a patchwork or a same-sex family) cannot be detected. The Finnish Immigration Service (2008; see also Chapter 3) acknowledges this: 'Purely biological relationship is not … sufficient for a positive decision on resident permit without the background of a genuine, permanent family life. A foster child, for example, could therefore be granted a resident permit on the basis of family ties if identifiable as an integral member of the family'.

How, then, can the pro and cons of DNA testing versus interviews be balanced in family reunification procedures? As this alternative only arises where children are involved, I will consider only these cases. The task of the family reunification procedure is to verify the existence of family ties and to discover possible cases of child trafficking. Human rights and ethical principles demand respect for autonomy, privacy and family life. Respect for autonomy requires that neither interviews nor DNA tests should take place under coercion. Given the special situation of family reunification procedures, in which a refusal to cooperate with the authorities (by not agreeing to either a DNA test or an interview) is bound to result in the rejection of the whole application, the entire process of family reunification is coercive by nature. In these circumstances true (informed) consent to be interviewed or to undergo a DNA test can never be achieved, as it can never be fully voluntary. Interviews and DNA tests thus seem to be equally problematic as far as individual autonomy and liberty are concerned.

With regard to privacy and respect for family life, the situation is different. In fact interviews, especially where small children are involved, are potentially more intrusive and distressing than DNA tests, always providing that the genetic data collected are used only in the specific reunification procedure and then destroyed together with all tissue samples. The fundamental problem of DNA testing in family reunification procedures therefore is not connected to privacy issues but to the fact that it does not verify family life, which is a social phenomenon and not a biological relation between DNA molecules. But as the existence of biological ties between a mother, a father and a child makes it at least more probable that there are also social bonds between them, a positive DNA test should be a sufficient indication for family ties. I would therefore argue in favour of initial DNA tests, which should be complemented by interviews only if their result is negative, in order to establish if there are real social family ties even if there is no biological kinship. Not DNA testing but interviews should be the last resort. This inverting of the current logic of family reunification procedures in all three countries examined, Austria, Finland and Germany, would not only minimise the ethical issues of the procedure but also shorten the entire process and free the applicants from the arbitrary decision of government officials, thus revealing the possible empowering aspects of this special kind of biological citizenship (see Chapter 6).

Chapter 6
Governing DNA Analysis for Family Reunification: A Comparative Perspective

Torsten Heinemann, Ilpo Helén, Thomas Lemke, Ursula Naue and Martin G. Weiss

This volume has analysed the complex historical and political trajectories of the use of DNA evidence for family reunification in Germany, Austria and Finland, and has also looked at ethical concerns related to this practice. This concluding chapter provides a detailed comparison of the rationalities and procedures involved in carrying out DNA testing for family reunification, and highlights important similarities as well as some remarkable differences between the three countries.

As DNA analysis for the 'verification' of family ties is a scientifically founded and well-established technology standardised by international laboratory accreditation systems (Morling et al. 2002; Gjertson et al. 2007), the technical procedure itself does not differ significantly in the three countries we investigated. In Germany and Finland (as in many other countries) only authorised and certified laboratories following certain protocols are allowed to perform DNA analysis for the verification of family ties. In Austria accreditation according to ISO 17025/2005, which guarantees the quality of forensic procedures, is not explicitly required. Nevertheless, most Austrian laboratories are certified by the German accreditation organisation and conduct DNA testing according to ISO norms as well as in accordance with the recommendations of the paternity testing commission of the International Society of Forensic Genetics (Morling et al. 2002; Gjertson et al. 2007), even if they are not in possession of an accreditation certificate. While the technical standardisation may be expected, our study also provides evidence that DNA analysis has become a routine practice in family reunification procedures in all three countries – a tendency that has been recently highlighted in other studies focussing on national contexts (Murdock 2008; Gomes-Cardoso 2009; Esbenshade 2010; La Spina 2012; Dove 2014; Lakhani and Timmermans 2014; Tutton, Hauskeller and Sturdy 2014). This development has far-reaching consequences, because family-related immigration is the statistically most significant mode of legal long-term immigration to the European Union (European Commission 2011; Müller 2012). Thus, any changes to legal frameworks and administrative practices for recognising family relations affect a large number of (potential) migrants and refugees.

Beyond stressing these common tendencies and shared paths of employing DNA analysis for family reunification, the most important result of our study is that the three countries compared in this volume represent very different regimes for the use of DNA evidence for verifying family relations in the context of immigration. Our comparative analysis shows that despite several attempts to harmonise immigration policies in Europe, the formal requirements and practical arrangements for family reunification in general, and for DNA testing in particular, vary significantly between European countries (Cholewinski 2002; Kofman 2004; Schibel 2004; Heinemann, Naue and Tapaninen 2013). Our study also portrays in detail the effects of routinisation of the use of DNA information in migration policies on those affected in the three countries under investigation. It highlights the different regimes of understanding and handling family reunification as such, alternative paths of implementing DNA analysis in this domain, and multiple ways of evaluating and assessing this kind of evidence. Our study thus contributes substantially to an empirically informed discussion and critical assessment of DNA kinship testing for family reunification within the European Union.

This concluding chapter begins with a comparison of the multiple legal definitions and formal regulations governing decision making on family reunification in the EU, in Germany, Austria and Finland. We then present the different regimes of implementing DNA evidence in the field of immigration and outline the practical arrangements for parental testing in family reunification. The third part of the chapter discusses the culture of mistrust reflected in the different administrative procedures and the employment of DNA evidence as a means to establish truth and trustworthiness. After that, we link our findings to the ongoing debate on biological citizenship and attempt to redirect the discussion on the basis of our empirical findings. The final section highlights some lessons learned from our comparative study and argues for a consistent, transparent and problem-oriented use of DNA analysis for family reunification, taking account of the shortcomings and problems we diagnose in current administrative practices.

Diversity of Legal Definitions and Formal Regulations

Although the EU introduced a common European Regulation for family reunification in 2003, the member states and in particular the three countries under investigation understand and regulate what they subsume under this label in very different ways. This begins with the definition of the family and the criteria for who is eligible to apply for family reunification. The Charter of Fundamental Rights of the European Union guarantees the right to have a family in Article 7 and 9 (see Chapter 1). However, beyond the general provisions of these articles, the Charter 'is silent on the question of family reunification and rather muted on the question of protecting the family' (Cholewinski 2002, 276). Actually, it does not even provide a definition of the family. Council Directive 2003/86/EC, the most important legal document on family reunification on the European level, is slightly more specific: 'Family reunification should apply in any case to members of the nuclear family, that is to say the spouse and the minor children'. Furthermore,

Article 10 of the Directive specifies that 'It is for the Member States to decide whether they wish to authorise family reunification for relatives in the direct ascending line, adult unmarried children, unmarried or registered partners as well as, in the event of a polygamous marriage, minor children of a further spouse and the sponsor'. In addition, Article 2 of Council Regulation No 343/2003, which complements Council Directive 2003/86/EC, clearly states that the family has to be understood in a broader sense than just the nuclear family.

However, lawmakers and regulators, particularly in Austria and Germany, enacted a narrower definition of the family. In both countries, only adult spouses and their underage children are eligible for family reunification, so that in fact the semantic field of 'family' is reduced to the nuclear family. It is also noteworthy that children who are above the age of 18 and siblings of any age do not count as family members in this legal understanding. Additionally, in Austria family reunification is only granted to family members who have had a common family history in the country of origin for at least two years. The legal understanding of the family in Austria and Germany is not, however, restricted to heterosexual married couples and their children, as registered same-sex partners are also eligible for family reunification.

In Finland, the legislation is slightly more liberal insofar as unmarried partners may also qualify as family members. Furthermore, underage siblings can apply to be reunited with their sisters and brothers, which is impossible in Germany and Austria. Finally, guardians can apply for family reunification with children they care for including foster children. However, it is apparent that in all three countries the right to family reunification is based on a very strict definition of the family that does not accommodate alternative family concepts in the countries of origin (for example multi-spousal families, or adults caring for younger relatives who are not their biological children, without the legal status of adoptees or foster children). It also contradicts, to some extent, the understanding of the family in the host countries. Moreover, the rise of divorce and remarriage and the growing recognition of same-sex unions in European countries have generated heterogeneous patterns of family structure which are not adequately addressed in the regulations governing family reunification in the three countries.

Fundamental differences characterise the formal requirements and institutional arrangements of the application process for family reunification, which also impacts on the use of DNA testing in this context. In the German context the application for family reunification is filed at the embassy, either by the sponsor from Germany or the family members who are seeking a residence permit. Normally, the applicants hire a lawyer in Germany who takes care of the application for them. The embassies are in charge of the entire procedure and will also make the final decision. They may contact the Aliens Department in the town or administrative district where the sponsor is living, to check whether the applicant can provide enough living space and a living wage for the prospective united family. Applicants have the right to appeal against the final decision taken by the embassy.

In Austria there are two separate procedures for family reunification. While the Federal Asylum Agency and the Residence and Citizenship Department are both subordinate authorities and parts of the Austrian Ministry of the Interior, they follow different application protocols. Both authorities cooperate with the embassies. The two procedures apply to different kinds of reuniting persons, that is to say 'sponsors', seeking to reunite with their family members and with particular legal statuses. The Settlement and Residence Act defines the legal category of the 'regular' immigrant and the requirements regarding income, housing and language skills as well as the administrative protocol for reuniting with his/her family. Refugees applying for family reunification are subjected to a different set of practices described in the Asylum Act and handled by the Federal Asylum Agency. This set of practices does not lay down any material requirements. Family reunification of refugees is framed as an asylum procedure in which the family member requests the same protection as was granted to the sponsor. This request has to be filed on Austrian territory. During this process, the refugees receive assistance from NGOs in tracing relatives in other countries, initiating the process of family reunification and communicating with the Federal Asylum Agency and the respective embassy. In order to enter Austrian territory, the refugee's family members have to apply for a visa at an embassy. The embassy will contact the Federal Asylum Agency and ask the agency to provide an opinion on the probability of the family members' receiving asylum in Austria. This prediction of the outcome of a putative asylum application is not considered a procedural step which the family can legally challenge, nor is it an actual decision. Only when the prediction is positive and the embassy has no further doubts about the identity of the family members are they allowed to enter Austria and apply for family asylum.

In Finland all family reunifications, irrespective of the legal status of the applicants, are managed by the Finnish Immigration Service (Migri). The procedure is informed by entrenched forms of cooperation between the Finnish Immigration Service, the police and municipal authorities in Finland, the embassies and missions abroad, and the authorised laboratory. As in Germany and Austria, in Finland recognised refugees wishing to reunite with their family do not have to meet any material requirements. This applies only if the family had been formed before the sponsor arrived in Finland (and excludes for example Finnish residents who marry foreigners). The most important difference is that the administrative system in Finland is much more centralised. In stark contrast, Germany has a decentralised system in which the embassies in the countries of origin, with the help of the Aliens Departments in the local districts and towns, have their say in the procedure. Austria can be described as a middle ground with its two distinct but equally relevant procedures for applying for family reunification.

Finally, there are remarkable differences in how Austria, Finland and Germany have implemented the use of DNA testing for family reunification. Finland was one of the first countries to regulate DNA testing in this domain by law with the amendment of the Aliens Act in 2000. Following a pilot project on the use of

DNA tests in family reunification cases from 1996 to 1999, the Finnish parliament proposed changes to the Aliens Act of 1991 by adding the detailed regulations for the use of DNA analysis for family reunification with an almost unanimous vote in favour (see Chapter 3). In Germany, authorities started to use parental tests for family reunification in 1991. In stark contrast to Finland, DNA analysis in this field was conducted without legal regulation for almost 20 years. Even today it is not part of the German Immigration Act. Instead, DNA testing for family reunification is addressed in the German Genetic Diagnostics Act, which came in force in February 2010. As we have demonstrated, the legal provisions are still insufficient and establish a double standard for German citizens and immigrants (see chapters 2 and 5). Austria was the last of the three countries to introduce DNA analysis for family reunification, and started to deploy this technology in the early 2000s. The use of DNA testing for family reunification of recognised refugees, which was already presented as an option that could provide 'proof' of family ties in the 2005 version of the Residence and Settlement Act, was introduced as a legal option in the 2009 amendment of the Austrian Aliens Law Package following a pilot project that lasted from 2006 to 2009 (see Chapter 4).

Heterogeneous Administrative Procedures and Practical Arrangements

The plural legal provisions and the diversity of formal regulations governing the field of family reunification in the three countries also result in discrepancies in the administrative procedures and incompatible practical arrangements for implementing DNA testing. This can be illustrated by considering the question of who is asked to provide DNA evidence in the countries investigated. German immigration officers in the Local Aliens Departments distinguish between two main groups of applicants for family reunification. Firstly, there are German nationals who marry a foreigner. This situation accounts for slightly more than 50 per cent of all applications for family reunification. In these cases, officers have to check if the requirements on living wage and living space for the united family are fulfilled. The second group concerns applications for family reunification by recognised refugees. Both cases follow the same legal procedure but with some alleviation for accepted asylum seekers and refugees, particularly by granting a three-month period after the asylum seeker has been officially recognised by the Federal Office for Migration and Refugees. During this timeframe, recognised refugees do not have to provide proof of secure income, living space or German language skills. It should be emphasised that German citizens may also be subjected to a DNA test, as officers generally consider the test as a further document that might supplement identity or marriage certificates deemed unreliable.[1] We are not

1 For example, this may happen if a German citizen is married to a foreigner and the family does not (yet) live in Germany. If the family consists of children, these children are by definition also German by birth. If they apply for a German passport, the officers in the

aware of any Austrian or Finnish citizens wanting to reunite with their foreign family members who were requested to provide DNA evidence. Thus, in Austria and Finland DNA analysis in the field of immigration seems to be conducted exclusively on refugees.

The material and practical constellations of providing DNA samples in the course of the application process also vary significantly in the three countries. The sponsor in Germany may or may not be a refugee, but for the purpose of family reunification everybody is subjected to the same conditions for DNA testing and finding legal support. All costs for the procedure have to be met by the applicant and his/her family. In Austria, on the other hand, the reuniting person requested to provide DNA evidence is mostly integrated into a philanthropic network of NGOs and some sort of community support that assists recognised refugees free of charge. In Finland, refugees and other Finnish citizens rely on the social welfare system of the country.

These three distinct constellations result in significantly different settings in which the whole procedure takes place. In the German context, a lawyer is almost always employed at the applicant's own expense, irrespective of whether the applicant is a refugee or a German citizen. The complexities of the application process are one reason why applicants are likely to seek legal counselling. The second reason, as some lawyers and representatives of German NGOs pointed out, is that German immigration officers sometimes delay the proceedings and lawyers can be helpful in keeping the procedure going or even speeding it up. At the same time lawyers are not allowed to help their clients to find a laboratory. In Germany, parental testing for family reunification is conceived as a competitive market where a situation in which a lawyer recommends a particular laboratory to their clients is seen as advertisement that is prohibited by the Federal Lawyers' Act (Bundesrechtsanwaltsordnung, BRAO). In line with the conception of a free market, there are more than 50 laboratories that offer the so-called immigration test for family reunification. Market access is only restricted by the requirement of the appropriate accreditation certificate. In fact, the price for a DNA test varies widely, from 300 to 1,500 euros for a family of three persons. On average, the test costs approximately 200 to 250 euros per person tested. It has to be paid for by the applicants irrespective of their legal status.

Finland can be seen as the opposite extreme to the German case with respect to market regulation. In Finland, the Hjelt Institute of the Department of Forensic Medicine at the University of Helsinki has been, in practice, solely responsible for analysing DNA samples taken in Finland or abroad in response to a request from Migri, and since 2014 it has been the only officially authorised laboratory for this purpose.[2] Thus, no market for DNA analysis exists in Finland. In addition,

German embassy may question the certificate of birth issued by the foreign country and request a DNA kinship test instead.

2 Before that date, the National Institute of Health and Welfare also had an authorisation.

the price is regulated by a government decree on forensic genetic paternity testing from 2005, and it is currently 269 euros per person for a DNA test. The Finnish Aliens Act provides payment for the tests from state funds, but the applicants have to reimburse the costs if they have found to have 'deliberately given false information' (see Chapter 3).[3]

The Austrian scenario lies in between the Finnish and the German variants. In Austria lawyers are generally not involved in applications for family reunification of refugees. Legal counselling and representation in court are provided by NGOs, free of charge. In cases where counsellors also act as legal representatives, they typically have a university degree in law. In contrast to German lawyers, the Austrian NGOs help applicants to find a laboratory for the DNA test. They generally recommend the laboratory nearest to the place of residence of the reuniting person, without comparing the prices charged by different laboratories. Even the immigration officers lack a complete overview of laboratories and prices. Therefore, there is no real competition between the four university-based laboratories and the three private laboratories conducting DNA tests for family reunification in Austria. The tests' pricing and payment modalities differ significantly, with the cheapest DNA tests costing between 506 and 550 euros for three persons tested (the charge for each additional person is 200 euros). 960 euros for three persons with 300 euros for each additional person is the middle of the price range. The highest price a laboratory charged was 600 euros for each person tested.

The payment policies in Finland and Austria can be seen as two contrasting methods for dealing with attempted fraud through monetary sanctions. In Finland the expenses are generally covered by the state but can be claimed back in cases where deliberate fraud is suspected, and in Austria it is exactly the other way round. Here the rationale behind the juridical and administrative logic is clearly based on suspected fraud, as applicants in principle have to pay for the DNA test and the costs will be refunded if the result affirms biological relatedness.

Another difference between the countries we investigated concerns the sources of evidence used to determine family relations. In Germany, the procedure is based on 'documentary evidence' alone. All proofs of an existing family relation have to be documented in one way or another, whether this is by documents of identification (for example ID cards and passports), certificates of birth, marriage or death, written statements based on investigations by lawyers officially accredited by the German embassies, or the report about the biological relation drawn up on the basis of a parental test. Therefore it is not 'genuine family life' that is the object of enquiry, but the documented, that is to say mostly the *contractual*

---

3 Until 2014 Migri could, in principle, accept the results of a DNA test arranged by the applicants themselves if the sample taking had been supervised by a Finnish public authority (for example, a police officer of an embassy official). Today, Migri does not take 'private' tests into consideration as a piece of evidence in any case (Finnish Immigration Service n.d.).

family relation. Biological parenthood verified by a DNA analysis serves as an indication for a partnership and in this sense substitutes the certificate of marriage or eliminates doubts (in cases of suspected fraud). In Austria, as in Germany, it is the claimed parental relationship that is investigated and not family life as such. Austrian authorities will conduct interviews with the applicants to complement and assess the truthfulness of documentary evidence and statements in the context of subsequent interviews during the asylum procedure.

In Finland, interviews are a key source of evidence. Migri also asks for documents, but even if no identity papers or marriage certificates can be provided an application procedure will be initiated on the basis of the interviews – which is impossible in Germany. As a consequence, Finnish authorities mainly investigate an existing social familial life rather than the documented family relations. This procedure may seem more humane or liberal, in the sense that it sets limits to a biological understanding of family ties. However, the investigation geared to prove the authenticity of family life can be more uncertain and more stressful for the applicants (see Chapter 5). Moreover, the results of DNA analysis can be used to question the credibility of the statements given by the applicants in interviews and not all applicants are offered DNA testing even if they request it (see Chapter 3).

Fundamental differences also characterise decision-making procedures in the three countries. In Germany, with a DNA test result the final decision is sometimes made within less than four months following the application. In Austria, after having received the results and the report of the DNA analysis stating the biological relation between the (presumed) family members, an officer in the Austrian Federal Asylum Agency takes a decision on whether the family members abroad get the entry visa or not. The agency then informs the embassy about the positive prognosis of decision and instructs it to issue the entry visa(s). In quite a lot of cases, it still takes many months to reach the final decision. In case of a positive outcome, family members must enter Austria within a period of six months. If this does not happen, the whole procedure has to start again with an evaluation of whether the applicants are still eligible for family reunification (for example regarding age). From the time the application is filed until the reunion with the family there is usually a period of six to twelve months, which can become even longer if problems arise.

In Finland, the processing times of applications for family reunification filed by refugees are noticeably longer than those filed by other third country nationals, and they clearly exceed the nine months set out in the Directive on family reunification and in the Aliens Act (Section 69 A). The average time varies between 15 months (for children) and 21 months (for a spouse or other family members). The congestion of applications and prolonged processing times are attributed to the absence of verifiable documents, which makes personal interviews and sometimes also DNA evidence necessary. The time needed for performing a DNA analysis is only two to three weeks, but it is prolonged considerably by the need for bureaucratic initiative, cooperation and decision making. Finally, and here we encounter another important difference to Germany and Austria, the test results are not given to the applicants themselves but to the Finnish Immigration Service.

Establishing Trustworthiness, Combating Fraud

DNA analysis for family reunification is far from being the only technological option applied in the government of migrants. Biotechnological testing and screening devices are increasingly being deployed in the current regime of migration control throughout Europe and particularly in the European Union, and fingerprinting and the related Eurodac database, national forensic databases and biometric residence cards are routinely used (Aas 2011, Dijstelbloem and Meijer 2011, Tutton, Hauskeller, and Sturdy 2014). Medical age assessment – bone X-rays which allow inferences of the likelihood that a person is still underage and therefore entitled to a more favourable treatment as an asylum seeker – is also incorporated into the policing of asylum and family reunification in many EU Member States, Germany, Austria and Finland included.

DNA testing as a routine procedure in governing migrants applying for family reunification, which we observe in Germany, Austria and Finland, manifests 'an obsessive deployment of surveillance strategies' (Fassin 2011), brought about by a restrictive turn in European immigration policies during the past two decades (European Commission 2011). In recent years, the rationale of fighting alleged fraud has become more predominant than before in border control and immigration policies in Europe. In the light of these developments, the migrants themselves and the evidence they might present become suspect to the authorities. In their eyes, the documents are likely to be unreliable and the narratives doubtful, while 'the body does not lie' (Aas 2006). Consequently, border control and immigration authorities increasingly deploy advanced biotechnological devices and forensic science, DNA testing included, to extend monitoring and supervising practices to the cellular and molecular levels of the 'alien' body.

Unlike DNA profiling in forensic contexts, the prime use of DNA kinship testing is not the identification of individuals but the estimation of genetic proximity (biological family relations). And it is the (non-)existence of these relations that is at stake in decision making on family reunification. DNA testing is clearly connected to the great concern about fraud, which is manifest in many recent European policy papers and reports on family reunification (European Commission 2011; Müller 2012; see also Chapter 3). In this environment of ubiquitous suspicion, DNA analysis stands out as the most exact technique for verifying both the family ties of the applicants and their narratives and testimonies. The immigration officials we interviewed in Germany, Austria and Finland shared this view of and trust in DNA analysis as the provider of uncomplicated 'truth' and supposedly failsafe evidence on which they could base their decisions on family reunification cases (Kruse 2010, 2012; Lynch, Cole and McNally 2008; Machado and Prainsack 2012).

So, if DNA testing is perceived as the solution, what is the question it purports to answer? In fact, the use of DNA evidence responds to rather different questions and concerns in the three countries. German immigration management focuses on 'verifying' biological links between family members. It minimises the need to evaluate complicated varieties of relatedness, rejects interview evidence and accepts a rather straightforward geneticisation of family ties (see Chapter 2). The primacy of biological kinship ties even has the effect that in exceptional cases polygamous families can be granted residence permits on the basis of DNA testing.[4]

By contrast, in Austria and Finland it is not the biological truth itself that is at the centre of administrative attention. Instead, the genetic 'facts' constitute means to measure and assess the applicant's credibility and truthfulness. The procedures in these countries put a lot of weight on the interviews, which makes the administrative reasoning more complicated and contingent compared to the German immigration regime since all pieces of evidence are weighed against one another, and even a positive result of a DNA test does not necessarily lead to a successful family reunification. In Austria, DNA testing is requested if the narrative evidence does not convince the authorities. In such cases, DNA testing is seen to establish the trustfulness of the refugees and the coherence of the story they have told. However, we encountered some cases in which the officer of the Federal Asylum Agency decided to go back to the documents and demanded certificates of marriage and birth even after the genetic relations were confirmed. This initiated a vicious circle for the applicants, as they could not provide any other documents than those which had resulted in the request for DNA evidence in the first place (see Chapters 3 and 4).

Finland is different from Austria in one important aspect: Migri offers the DNA test to the applicant without valid documents if the narratives and other investigations provide sufficient evidence of 'actual' family ties. Here, the administrative attention focuses on establishing an existing family life. As a Finnish immigration official said, 'there has to be a family to reunify' (Finnish official 4). This clearly limits the significance of DNA testing compared to other sources of evidence. A Migri press release in 2008 clearly stated that DNA evidence is not prioritised over other pieces of evidence: 'A purely biological relationship is not, however, sufficient for a positive decision on a residence permit without a background of genuine, permanent family life' (Finnish Immigration Service 2008).

4 We know of one case of family reunification where a refugee from Somalia claimed to be married to two women, which is not allowed in the European Union. As the man had children with both women, which was proven by a parental test, he could bring all his biological children to Germany and the mothers could then apply to be reunited with their biological children. In the end, all the persons that constituted the family were allowed to enter Germany. This would have been much more difficult or even impossible without the DNA test (Applicant 5).

This reservation illustrates the paradoxes of the Finnish procedure: DNA testing has become routine, yet the genetic conception of family ties is played down in rhetoric and in decision making. It allows the rejection of an application on the basis of insufficient proof of family life despite the positive result of a DNA test, that is to say the affirmation of biological family bonds. In such cases, the migration officials may take into consideration, for example, the reasons why the family was separated in the first place, continuity of and commitment to family ties despite the separation, and 'dependency' or 'the best interest of the child', and set these considerations against the DNA evidence. Some applications have been rejected despite a positive DNA result because the sponsor had given false information on his or her family ties at the beginning of the process, or because the family tie had ceased to exist during the long separation. By contrast, narrative and other evidence of family life may also outweigh a negative result of DNA analysis; our material shows that this is most likely to happen in cases involving foster children.

Thus, in all countries we encountered procedures and administrative patterns that can be described as inconsistent, contradictory or even paradoxical. They seem to be determined and shaped by the prevailing political aim to reduce the number of successful applications for family reunification. We also found that immigration authorities rely on quite different and heterogeneous definitions of family across and within the three countries. In some cases, family is referred to as a marital and therefore primarily contractual relation, in which children are by definition the children of the spouses, while on other occasions it is the other way round and the fact that two persons are the biological parents of a child makes them a family that is eligible for family reunification. This observation is also true for the ambivalent status of identity papers in the administrative process. In Germany, Austria and (to a lesser extent) Finland, the process of family reunification does not begin until the applicants present travel documents, particularly a passport and documents supporting their legal residence in the country where the embassy is located. This request gives rise to an ambivalent situation, because these documents are often considered invalid or not suitable according to European regulations and requirements. Thus, identity papers are formally required to initiate the application process yet are immediately rejected by the authorities. This holds true for example for Somali passports in Finland, for passports from certain 'failed' or instable states in Austria, and for identity papers from blacklisted countries in Germany.

Current immigration policy in Europe is characterised by suspicion and distrust towards migrant 'aliens'. In this context immigration authorities see DNA analysis as an efficient and objective tool, which can provide scientific and exact information and thus facilitate and improve decision making. The immigration officials consider DNA evidence exact and beyond manipulation as compared to documents and applicants' narratives. However, our study shows that DNA analysis per se – besides the fact that it is restricted to family members connected by genetic ties – rarely provides the final and definitive answer in family reunification cases. Especially in Austria and Finland, the 'truth' of genetics is assessed in connection with other pieces of evidence, quite similarly to what Lynch, Cole and McNally (2008) or Heinemann, Lemke and Prainsack (2012) have described for the forensic context. In the end the whole procedure is about credibility, and it is the trustworthiness of the applicants that is investigated in documentary evidence, interrogated in interviews and tested in laboratories (see Chapters 3 and 4; also Aas 2006; Lynch, Cole and McNally 2008; Machado and Prainsack 2012).

Towards a More Complex Understanding of Biological Citizenship

The empirical findings of our study might also be fruitfully related to the current debate on the concept of 'biological' (Petryna 2002; Rose and Novas 2005) or 'genetic citizenship' (Kerr 2003; Heath, Rapp, and Taussig 2004).[5] This term has been described as a 'keyword in the making' (Cooter 2008, 1,725), and has gained much currency among scholars of science and technology and social scientists working on the societal and political implications of biotechnologies and biomedicine.[6] Since it was first introduced in 2002 by anthropologist Adriana Petryna (2002) in her study on Ukrainian citizens who were affected by the Chernobyl nuclear disaster of 1986, the growing literature on this topic has almost exclusively referred to the importance of patients associations, disease advocacy organisations and self-help groups. The claim is that they give rise to new forms of subjectivation and collective action, thus challenging existing borderlines between laypeople and scientific experts, between active researchers and passive beneficiaries of technological progress. Frequently, authors using the notion of biological or genetic citizenship argue that these groups are questioning access to knowledge and claims to expertise, forging new alliances with biomedical researchers, and lobbying to influence political decision making and to receive funding for medical research (Heath, Rapp and Taussig 2004; Rose and Novas 2005; Callon and Rabeharisoa 2007; Rose 2007; Fitzgerald 2008; Schaffer, Kuczynski and Skinner 2008; Fraser 2010; Reubi 2010).

So far, the focus of the debate on biological citizenship has been on the extension of rights, the emergence of new possibilities of civic participation and social engagement, and on the choice-enhancing options of biomedical technologies and especially the new genetics. One of the most prominent examples of this reading of biological citizenship is an article by Nikolas Rose and Carlos Novas (Rose and Novas 2005; Rose 2007, 131–54). The authors diagnose a 'new

5 The following arguments are developed in more detail in Heinemann and Lemke (2014) and Helén (2014).

6 Following the work of Thomas H. Marshall (1950), there is an ongoing scientific debate stressing the decomposition and extension of citizenship beyond formal membership in a state into new domains. Closely related to notions like 'economic' or 'social citizenship', research on 'biological citizenship' focuses on claims about benefits from social policies or economic participation (Fahrmeir 2007; Isin, Nyers and Turner 2008).

form of citizenship' (Rose and Novas 2005, 439) in Western democracies, and see in practices of biological citizenship an important challenge for the nation-state and its forms of regulation. Rose and Novas distinguish between two dimensions of this process. Firstly, they link biological citizenship to hitherto unknown forms of participation and legal claims that question traditional forms of nationally bound citizenship rights. According to Rose and Novas, bioscientific knowledge and biotechnological innovations transgress national boundaries and regulatory competences. Similarly, patient organisations and support groups often choose transnational forms of organisation, and the internet is becoming more and more important as a means of information and communication platform for patients and people with disease risks (see also Schaffer, Kuczynski and Skinner 2008). Secondly, the authors combine their observation of a pluralisation and fragmentation of traditional political spaces with the thesis that there has been a historical break; they argue that biological ideas and prejudices have always shaped concepts of citizenship. By implicitly or explicitly specifying rules of membership, conditions of participation and criteria of access, such ideas defined who and on what biological grounds could be a candidate for citizenship (on the basis of sex, race, etc.). However, Rose and Novas insist that there is a decisive rupture between the eugenic projects and racialised politics of the past and the new genetics. They argue that biological citizenship stands for a new governmental regime which radically breaks with the eugenic and racialised past, and that there has been a shift from political rationalities directed toward the management of risk at the level of populations to the individual management of genetic risks (Rose and Novas 2005, 132–33; Rose 2007, 54–64). Thus, Rose (2007, 132) claims that 'contemporary biological citizenship, in the advanced-liberal democracies of "the West" ... does not take this racialized and nationalized form'.

While this reading of biological citizenship certainly highlights important social and political implications of biotechnological innovations, it tends to downplay or ignore practices of surveillance and exclusion and the refusal of citizenship rights based on biological knowledge (Amoore 2006; Lemke and Wehling 2009; Raman and Tutton 2010; Brekke and Sirnes 2011; Wehling 2011).

On the basis of our empirical findings as presented in this study, we suggest a richer and more comprehensive account of biological citizenship that overcomes the blind-spots and limitations of the current debate, with particular reference to two analytical problems. First, the literature on biological citizenship (too) often stresses the transnational dynamics of patient organisations and support groups. We do not deny the medical significance and the societal impact of support group and patients associations for (genetic) diseases that are characterised by forms of organisation and modes of communication often transgressing national borders (see also Epstein 1996; Callon and Rabeharisoa 1999; Brown et al. 2008). However, the use of DNA tests in immigration displays the enduring relevance of biological criteria to determine who should be granted citizenship rights in a particular nation-state. It follows that there is not only a transnational dynamic to be observed but also a continuation and re-articulation of the relation between biology and citizenship. This also means that the second claim of a historical break with past population policies has to be reconsidered. Biological citizenship refers not only to a displacement towards the individual management of genetic risks, but also to a form of migration control targeted at particular populations – in Germany especially those who derive from 'blacklisted countries', mostly from sub-Saharan Africa and Central and South East Asia, while in Finland Somalis are singled out as the target of DNA analysis (see chapters 2 and 3). In this respect, the thesis that the use of genetic information for population control belongs to the biopolitics of the past has to be complemented or even corrected. Thus, immigration management and border control are central domains where the 'eclipsed' side of current biological citizenship materialises (Muller 2004; Aas 2011; Ticktin 2011; Nayar 2012; Heinemann and Lemke 2014; Helén 2014; Lakhani and Timmermans 2014).[7]

The use of DNA tests for decision making in the context of immigration reveals the selective format of the debate on biological citizenship. So far, this discussion has often stressed the biological body as the basis of claims about social inclusion, recognition and democratic deliberation. However, as our study demonstrates, this view of biological citizenship itself excludes important dimensions of contemporary migration regimes. For example, the use of DNA testing in immigration policies does not signify the advent of a 'molecular biopolitics' (Rose 2007, 12) that finally displaces the concern with bodily features such as skin colour, hair texture or eye shape. Rather, it serves to reaffirm and re-establish 'traditional' forms of classification and exclusion. One central question that Rose and Novas point to (2005, 440), and which has been neglected so far in the literature on biological citizenship, addresses the link between biology and citizenship, asking what biological prerequisites an individual must fulfil in order to be recognised as a citizen. In decision making on family reunification biological features, in a very literal sense, have become the basis for citizenship claims. To be sure, it is not the presence or absence of isolated biological traits of an individual or group that is at stake in this case but rather biological relations between individuals as family members. But there is a strong connection between biology and citizenship rights for at least two reasons. DNA analysis in the context of immigration first affirms and institutionalises a genetic concept of family. Second, it is mainly certain quite specific groups who are subjected to this procedure in the first place, which may be interpreted as racial profiling. By showing that population-centred policies do not belong to the past but still inform contemporary biopolitics, the investigation

---

7 As long as 10 years ago, Anne Kerr (Kerr 2003, 44) rightly cautioned against 'the prevailing emphasis on transformation' in the debate on biological and genetic citizenship: 'Although it is important to reject simplistic parallels between the eugenics of the past and the genetics of the present, there is a danger that this focus on transformations is resulting in less attention being paid to the static and reactionary aspects of the 'new' genetics and its wider social context'.

of the use of DNA analysis in immigration decisions opens up a new dimension in the discussion of biological citizenship.

There is an additional aspect that has to be taken into account. In fact, there is a transnational dynamics of biological citizenship within immigration politics and management to be observed, but it is very different from the one connected with cross-border activism of patients' networks. The literature on biological citizenship focuses on the *extension* of rights, participation and 'empowerment' of engaged biocitizens, while appeal to biological relations and to information provided by DNA analysis in immigration management concerns the problem of *acquiring* a legal status and (some) rights as an immigrant or refugee. In the latter terrain, the mode of national state citizenship, with its exclusive elements, is challenged by post-national citizenship based on internationally recognised human rights of migrants. In the procedure of family reunification, DNA information on parenthood may strengthen the personal rights of the asylum seeker or immigrant based upon universal fundamental rights, since the biological family tie provides the basis for the right to enter the country and stay there (Helén 2014). Thus, DNA analysis connects immigrants' rights to biological information (see Chapter 3) and, consequently, 'biology' plays an important part in contestations between conventional constellations and traditions of national citizenship on the one hand and post-national citizenship claims in the field of immigration policy on the other.

Some Lessons Learned

In this final section, we highlight another important result of our study. The comparative analysis we have conducted not only provides valuable insights into the specific paths and problems of the immigration management in three European countries Austria, Finland and Germany, but also points to some important issues with regard to DNA analysis for family reunification that should be addressed in future regulatory practices. These issues relate to the application of DNA analyses and its consequences for administrative decision making.

Our findings stress the importance of increasing consistency and transparency in overall procedures and administrative practices. Especially in Austria, with its two different procedures and the involvement of a variety of governmental and non-governmental actors, the steps and administrative decisions taken in the course of the procedure surrounding DNA analysis for family reunification are often opaque. Similarly in Germany's highly decentralised administration, the legal regulations for family reunification are interpreted quite differently by the embassies and Aliens Departments, thus resulting in contradictory decisions in comparable cases. The need for transparency extends to the assessment of evidence. In Austria, we noted in some cases a vicious circle of first requesting documents and then DNA evidence, then demanding documents once again after the positive results of the DNA kinship test were provided. This might lead to a rejection of family reunification even though the DNA analysis has confirmed the biological ties between the family members, as they cannot provide the documents in question (see Chapter 4). To ensure transparency and reliability the results of the DNA test have to be accepted by the authorities once the option is offered to the applicants. This also implies that voluntariness as well as informed consent have to be guaranteed (see Chapter 5). As we have shown in detail, 'principles of free and informed consent are problematic in the case of people in dire situations with limited options and little knowledge of the risks involved, especially concerning unexpected results or potential use of the data for other purposes in the future' (Heinemann, Naue and Tapaninen 2013, 201).

Another important aspect that should be considered concerns data protection and storage. In contrast to the Finnish practice where the destruction of samples and data from the test after the decision is required by law (Heinemann, Naue and Tapaninen 2013, 201) the handling of generated data in Germany seems particularly problematic. Here the applicants' DNA profiles are stored for possible future use in forensic contexts, leading to a general criminalisation of the applicants (see also Aas and Bosworth 2013). To avoid this obvious stigmatisation and discrimination of the (mostly non-German) applicants, all data (and samples) should be used only for the specific purpose of kinship testing and then be destroyed (see Chapter 5).

It should be noted that DNA analyses for family reunification are often embedded in legal, social, and even racial discriminatory practices which are not only the result of how the techniques and procedures are employed – as in the German case, where the tested applicants are automatically stigmatised as possible criminals – but are often the very condition for the employment of DNA testing in the first place. The study provides empirical evidence that in all three countries DNA analysis is used most frequently with regard to members of specific ethnic and already racialised groups: Somalis in Finland and sub-Saharan Africans in Austria and Germany. There may be different and contingent reasons for the over-representation of these groups in DNA testing for family reunification, but the fact that these groups are tested more than others contributes to singling them out and racialising them further.

Keeping in mind the Universal Declaration of Human Rights, which states that family life is worthy of protection 'by society and the State', the right to family life should be the basis for decision making in the context of family reunification. The present comparative study contributes empirically to the ongoing discussion about harmonisation and standardisation efforts regarding family reunification policies within the EU. By analysing and confronting administrative practices with institutional settings and legal frameworks, it highlights the multiple and diverse effects of DNA analyses in the context of immigration and how they narrow down opportunities for people to reunite as a family when they have been displaced in their country of origin in the context of war, conflict or other migration-inducing situations. The study calls for a more empirically informed and problem-oriented perspective on the issue of DNA analysis and family reunification. This perspective emphasises the safeguarding of meaningful informed consent, the guarantee of voluntariness and the material accessibility of DNA analyses for applicants (the financial capacity to pay for them) on the one hand, and a structured, transparent and comprehensible administrative procedure including a more balanced account of evaluating and assessing different forms of evidence on the other. This approach should also be aware of the complex and diverse forms of family life in contemporary societies and inform further harmonisation and standardisation policies regulating family reunification within the EU and beyond.

References

Aallas, E., 1991. *Somalishokki*. Helsinki: Suomen Pakolaisapu.

Aas, K.F., 2006. 'The body does not lie': Identity, risk and trust in technoculture. *Crime Media Culture*, 2(2), 143–58.

Aas, K.F., 2011. 'Crimmigrant' bodies and bona fide travelers: Surveillance, citizenship and global governance. *Theoretical Criminology*, 15(3), 331–46.

Aas, K.F. and Bosworth, Mary (eds), 2013. *The Borders of Punishment: Migration, Citizenship and Social Exclusion*. Oxford: Oxford University Press.

Agamben, G., 1998. *Homo Sacer: Sovereign Power and Bare Life*. Stanford, CA: Stanford University Press.

Ahrens v. *Germany*, 2012. Application no 45071/09.

Allen, A.L., 1997. Genetic privacy: Emerging concepts and values. In: M.A. Rothstein, ed. 1997. *Genetic Secrets: Protecting Privacy and Confidentiality in the Genetic Era*. New Haven, CT: Yale University Press, 31–59.

Amoore, L., 2006. Biometric borders: Governing mobilities in the war on terror. *Political Geography*, 25(3), 336–51.

Andorno, R., 2004. The right not to know: An autonomy based approach. *Journal of Medical Ethics*, 30, 435–40.

Archard, D.W., 2011. Children's rights. In: E.N. Zalta, ed. *Stanford Encyclopedia of Philosophy*, [Online] Available at: http://plato.stanford.edu/entries/rights-children/ [accessed 24 April 2014].

Armstrong, D., Michie, S. and Marteau, T., 1998. Revealed identity: A study of the process of genetic counselling. *Social Science and Medicine*, 47(11), 1653–8.

Bade, K.J. and Anderson, L., 1994. Immigration and social peace in united Germany. *Daedalus*, 123(1), 85–106.

Bainham, A., 2008. Arguments about parentage. *The Cambridge Law Journal*, 67(2), 322–51.

Beauchamp, T.L. and Childress, J.F., 1994. *Principles of Biomedical Ethics*. New York, NY: Oxford University Press.

Beck, U., 1990. Elternschaft als Baukastensystem: Die genetische Selbstkorrektur und Selbstgestaltung der Menschennatur. In: U. Beck and E. Beck-Gernsheim. *Das ganz normale Chaos der Liebe*. Frankfurt am Main: Suhrkamp, 209–13.

Beck-Gernsheim, E., 1998. On the way to a post-familial family: From a community of need to elective affinities. *Theory, Culture and Society*, 15(3), 53–70.

Berger, Peter L. and Berger, B., 1983. *The War Over the Family: Capturing the Middle Ground*. New York, NY: Anchor Press.

Birks, M. and Mills, J., 2011. *Grounded Theory: A Practical Guide*. Thousand Oaks, CA: Sage.

Bledsow, C.H. and Snow, P., 2011. Family reunification ideals and the practices of transnational reproductive life among Africans in Europe. In C.H. Browner and C.F. Sargent, eds. *Reproduction, Globalization, and the State: New Theoretical and Ethnographic Perspectives*. Durham: Duke University Press, 127–91.

Blouin, M. S., 2003. DNA-based methods for pedigree reconstruction and kinship analysis in natural populations. *Trends in Ecology and Evolution*, 18(10), 503–11.

Bloustein, E., 1964. Privacy as an aspect of human dignity: An answer to Dean Prosser. *New York University Law Review*, 39, 962–1007.

Bogner, A., Littig, B. and Menz, W. (eds), 2009. *Interviewing Experts*. Basingstoke and New York, NY: Palgrave Macmillan.

Bohmer, C. and Shuman, A., 2008. *Rejecting Refugees: Political Asylum in the 21st Century*. London: Routledge.

Böckenförde, E.-W., 1976. *Staat, Gesellschaft, Freiheit.* Frankfurt am Main: Suhrkamp.

Brekke, O.A. and Sirnes, T., 2011. Biosociality, biocitizenship and the new regime of hope and despair: Interpreting 'Portraits of Hope' and the 'Mehmet Case'. *New Genetics and Society*, 30(4), 347–74.

Brown, P., Zavestoski, S., McCormick, S., Mayer, B., Morello-Frosch, R., and Altman, R., 2008. Embodied health movements: New approaches to social movements in health. In: P. Brown, ed. *Perspectives in Medical Sociology*. Long Grove, IL: Waveland Press, 521–38.

Bröckling, U., Krasmann, S. and Lemke, T. (eds), 2011. *Governmentality*. New York, NY: Routledge.

Bunn, G.C., 2012. *The Truth Machine: A Social History of the Lie Detector*. Baltimore, MD: Johns Hopkins University Press.

Butler, J., 2006. *Gender Trouble: Feminism and the Subversion of Identity.* New York, NY: Routledge.

Callon, M. and Rabeharisoa, V., 1999. La leçon d'humanité de Gino. *Réseaux*, 17(95), 197–233.

Callon, M. and Rabeharisoa, V., 2007. The growing engagement of emergent concerned groups in political and economic life: Lessons from the French Association of Neuromuscular Disease Patients. *Science, Technology and Human Values*, 33(2), 230–61.

Chambers, D. 2012. *A Sociology of Family Life: Change and Diversity in Intimate Relations*. Cambridge, UK and Malden, MA: Polity Press.

Chadwick, R.F., Levitt, M. and Shickle, D. (eds), 1997. *The Right to Know and the Right Not to Know.* Aldershot: Ashgate.

Cholewinski, R., 2002. Family reunification and conditions places on family members: Dismantling a fundamental human right. *European Journal of Migration and Law*, 4(3), 271–90.

Clarke, A., 2005. *Situational Analysis*. Thousand Oaks, CA: Sage.

Cole, S. A., 2001. *Suspect Identities: A History of Fingerprinting and Criminal Identification.* Cambridge, MA: Harvard University Press.

Cooley, T.M., 1880. *A Treatise on the Law of Torts or the Wrongs That Arise Independent From Contract.* Chicago, IL: Callaghan.

Cooter, R., 2008. Biocitizenship. *The Lancet*, 372(9651), 1725.

Council of the European Union, 2003. Council Directive No 2003/86/EC. *EUR-Lex, Official Journal of the European Union*, L, 12–18.

Crawley, H., 2012. *Working With Children and Young People Subject to Immigration Control: Guidelines for Best Practice.* Second Edition. London: Immigration Law Practitioners' Association.

Davis, A., 1994. Are you my mother? The scientific and legal validity of conventional blood testing and DNA fingerprinting to establish proof of parentage in immigration cases. *Brigham Young University Law Review*, 129–46.

Dawid, A.P., Mortera, J. and Pascali, V.L., 2001. Non-fatherhood or mutation? A probabilistic approach to parental exclusion in paternity testing. *Forensic Science International*, 124(1), 55–61.

Dean, M., 2010. *Governmentality: Power and Rule in Modern Society.* Second Edition. London: Sage.

Degtyareva, V., 2011. Defining family in immigration law: Accounting for nontraditional families in citizenship by descent. *Yale Law Journal*, 120(4), 862–909.

Diakonie Österreich, 2005. Stellungnahme der Diakonie Österreich zum Bundesgesetz, mit dem das Asylgesetz 2005 und das Fremdenpolizeigesetz 2005 erlassen, sowie das Bundesbetreuungsgesetz, das Personenstandsgesetz, das UBASG und das EGVG geändert warden, [Online]. Available at: http://fluechtlingsdienst.diakonie.at/Data/content/MediaDB/content/FD/content/downloadable-files-rechtliches/1142510470_2ot2vmp6dr/Diakonie%20Begutachtung%20AsylG%2005.pdf [accessed 24 April 2014].

Dickinson, B.D., Genel, M., Robinowitz, C.B., Turner, P.L. and Woods, G.L., 2002. Gender verification of female Olympic athletes. *Medicine and Science in Sports and Exercise*, 34(10), 1539–42.

Dijstelbloem, H. and Meijer, A. (eds), 2011. *Migration and the New Technological Borders of Europe*. Basingstoke and New York, NY: Palgrave Macmillan.

Dove, E.S., 2014. Back to blood: The sociopolitics and law of compulsory DNA testing of refugees. *University of Massachusetts Law Review*, 8(2), 466–530.

Edwards, J. and Salazar, C., 2009. *European Kinship in the Age of Biotechnology*. New York, NY: Berghahn Books.

Epstein, S., 1996. *Impure Science: Aids, Activism, and the Politics of Knowledge*. Berkeley, CA and London: University of California Press.

Esbenshade, J., 2010. *An Assessment of DNA Testing for African Refugees*. Washington, DC: Immigration Policy Center.

EUMC (European Monitoring Center on Racism and Xenophobia), 2000. *Diversity and Quality for Europe. Annual Report 2000*, [Online] Available at: http://fra.europa.eu/sites/default/files/fra_uploads/133-AR00_p2_EN.pdf [accessed 24 April 2014].

European Commission, 2011. Green paper on the right to family reunification of third-country nationals living in the European Union (Directive 2003/86/EC). *EUR-Lex. Official Journal of the European Union*, 735, 1–12.

European Migration Network, 2008. *Family Reunification: Synthesis Report*, [Online]. Available at: http://emn.intrasoft-intl.com/html/index.html [accessed 24 April 2014].

European Migration Network. 2009. *Ad-Hoc Query on Conducting Other Investigations (Using a DNA Test) in Family Reunification* [Online] Available at: http://emn.intrasoft-intl.com/html/index.html [accessed 24 April 2014].

European Migration Network, 2010a. *Ad-Hoc Query on the Concept of Family Member*, [Online] Available at: http://emn.intrasoft-intl.com/html/index.html [accessed 24 April 2014].

European Migration Network, 2010b. *Ad-Hoc Query on Verification of Legality and Genuineness of Marriage and Validation of Paternity*, [Online] Available at: http://emn.intrasoft-intl.com/html/index.html [accessed 24 April 2014].

European Migration Network, 2011. *Key EU Migratory Statistics*, [Online] Available at: http://emn.intrasoft-intl.com/html/index.html [accessed 24 April 2014].

European Migration Network, 2012a. *Misuse of the Right to Family Reunification: Marriages of Convenience and False Declarations of Parenthood*, [Online] Available at: http://emn.intrasoft-intl.com/html/index.html [accessed 24 April 2014].

European Migration Network, 2012b. *Misuse of the Right to Family Reunification: Marriages of Convenience and False Declarations of Parenthood – National Contribution from Finland*, [Online] Available at: http://emn.intrasoft-intl.com/html/index.html [accessed 24 April 2014].

Eurostat, 2011. *News Release 105/2011. Foreign Citizens Made Up to 6.5 Per Cent of the EU27 Population in 2010*, [Online] Available at: http://europa.eu/rapid/press-release_STAT-11–105_en.htm [accessed 24 April 2014].

Fahrmeir, A., 2007. *Citizenship: The Rise and Fall of a Modern Concept*. New Haven, CT: Yale University Press.

Fassin, D., 2005. Compassion and repression: The moral economy of immigration policies in France. *Cultural Anthropology*, 20(3), 362–87.

Fassin, D., 2011. *Humanitarian Reason: A Moral History of the Present Times*. Berkeley, CA: University of California Press.

Fassin, D. and d'Halluin, E., 2005. The truth from the body: Medical certificates as ultimate evidence for asylum seekers. *American Anthropologist*, 107(4), 597–608.

Federal Constitutional Court, 2009. *Pressemitteilung Nr. 98/2009 vom 25 August 2009: Richtervorlage zu § 9 Abs. 7 LPartG unzulässig*, [Online] Available at: https://www.bundesverfassungsgericht.de/pressemitteilungen/bvg09–098.html [accessed 24 April 2014].

Federal Court of Justice, 2008. *Decision of 30 July 2008 – XII ZR 150/ 06.*

Federal Government of Germany, 2008. DNS-Abstammungsgutachten im Rahmen von aufenthaltsrechtlichen Identitätsfeststellungen. Antwort auf die Kleine Anfrage der Abgeordneten Ulla Jelpke, Sevim Dagdelen, Jan Korte und der Fraktion DIE LINKE. *Drucksache 16/7698*, 1–5.

Federal Government of Germany, 2010. Vorwürfe gegen Ausländerbehörden wegen Gentests bei binationalen Eltern. Antwort der Bundesregierung auf die Kleine Anfrage der Abgeordneten Sevim Dagdelen, Ulla Jelpke, Wolfgang Neskovic, weiterer Abgeordneter und der Fraktion DIE LINKE. *Drucksache 17/1096*, 1–9.

Federal Office for Migration and Refugees, 2011. *Das Bundesamt in Zahlen 2010: Asyl, Migration, ausländische Bevölkerung und Integration.* Nuremberg: Author.

Federal Office for Migration and Refugees, 2013. *Das Bundesamt in Zahlen 2012: Asyl, Migration und Integration.* Nuremberg: Author.

Fehrenbacher, A., 2009. Das Anfechtungsrecht der Behörde bei rechtmissbräuchlichen Vaterschaftsanerkennungen zur Erlangung eines Aufenthaltstitels. *Zeitschrift für Ausländerrecht und Ausländerpolitik*, 29(1), 22–7.

Feller, E., Türk, V. and Nicholson, F. (eds), 2003. *Refugee Protection in International Law. UNHCR's Global Consultations on International Protection.* Cambridge, UK and New York, NY: Cambridge University Press.

Finnish Immigration Service, 2008. *DNA Analysis Has Unified a Record Number of Immigrant Families*, [Online] Available at: http://www.migri.fi/for_the_media/bulletins/press_releases/press_releases/1/0/dna_testing_has_united_a_record_number_of_immigrant_families_13968 [accessed 24 April 2014].

Finnish Immigration Service, 2014. *DNA Test in Connection With a Residence Permit Application*, [Online] Available at: http://www.migri.fi/download/33598_dna_eng.pdf [accessed 24 April 2014].

Finnish Immigration Service, (n.d.). *Absence of Travel Document*, [Online] Available at: http://www.migri.fi/moving_to_finland_to_be_with_a_family_member/submitting_the_application/absence_of_travel_document [accessed 24 April 2014].

Finkler, K., 2000. *Experiencing the New Genetics: Family and Kinship on the Medical Frontier*. Philadelphia, PA: University of Pennsylvania Press.

Finkler, K., Skrzynia, C. and Evans, J.P., 2003. The new genetics and its consequences for family, kinship, medicine and medical genetics. *Social Science and Medicine*, 57(3), 403–12.

Fischer, F., 2003. *Reframing Public Policy: Discursive Politics and Deliberative Practices*. Oxford and New York, NY: Oxford University Press.

Fitzgerald, J., 1998. Geneticizing disability: The Human Genome Project and the commodification of self. *Issues in Law and Medicine*, 14(2), 147–63.

Fitzgerald, R., 2008. Biological citizenship at the periphery: Parenting children with genetic disorders. *New Genetics and Society*, 27(3), 251–66.

Flick, U., 2007. *Designing Qualitative Research.* London: Sage.

Flick, U., 2009. *An Introduction to Qualitative Research*. London: Sage.

Fonseca, C.L.W., 2011. Following the path of DNA paternity tests in Brazil. In: C.H. Browner and C.F., eds. Sargent *Reproduction, Globalization, and the State: New Theoretical and Ethnographic Perspectives*. Durham, NC: Duke University Press, 138–53.

Foucault, M., 2000. Truth and power. In: M. Foucault and J.D. Faubion, eds. *Power*. New York, NY: New Press, 111–33.

FPÖ (Freiheitliche Partei Österreichs), 2007. Weinzinger. FPÖ fordert DNA-Analyse bei Familiennachzug, [Online] Available at: http://www.ots.at/presseaussendung/OTS_20071011_OTS0232/weinzinger-fpoe-fordert-dna-analyse-bei-familiennachzug [accessed 24 April 2014].

Franklin, S., 2000. Life itself: Global nature and the genetic imaginary. In: S. Franklin, C. Lury and J. Stacey, eds. *Global Nature, Global Culture*. London: Sage, 188–227.

Franßen-de la Cerda, B., 2010. Die Verpflichtung des Ausländers zur Mitwirkung (§ 82 AufenthG). *Zeitschrift für Ausländerrecht und Ausländerpolitik*, 30(3), 81–90.

Fraser, S., 2010. Hepatitis C and the limits of medicalisation and biological citizenship for people who inject drugs. *Addiction Research and Theory*, 18(5), 544–56.

Freeman, T. and Richards, M., 2006. DNA testing and kinship: Paternity genealogy and the search for the 'truth' of our genetic origins. In: F. Ebtehaj, B. Lindley and M. Richards, eds. *Kinship Matters,*. Oxford and Portland, OR: Hart Publishing, 67–95.

Frenz, W., 2008. Begrenzung ausländerrechtlicher Maßnahmen durch europäische Grundrechte. *Zeitschrift für Ausländerrecht und Ausländerpolitik*, 28(11/12), 385–8.

Fried, C., 1970. *An Anatomy of Values*. Cambridge, MA: Harvard University Press.

Funk, V., 2007. Ausländerbehörden setzen auf Gentests. *Frankfurter Rundschau*, 12 November, 6.

Funk, V., 2008. Mal DNA-Test, mal Familienfoto. *Frankfurter Rundschau*, 3 September, 9.

Funk, V., 2009. DNA-Test für Flüchtlinge. *Frankfurter Rundschau*, [Online 9 March] Available at: http://www.fr-online.de/home/dna-test-fuer-fluechtlinge,1472778,3347042.html [accessed 24 April 2014].

Garner, S., 2007. The European Union and the racialization of immigration, 1985–2006. *Race and Ethnicity: Multidisciplinary Global Perspectives*, 1(1), 61–87.

Gaserow, V., 2007. Grenzen dicht für Birmaner. *Frankfurter Rundschau*, 22 October, 8.

Gendiagnostikkommission, 2011. Richtlinien der Gendiagnostikkommission zu den Anforderungen an die Inhalte der Aufklärung gemäß § 23 Abs. 2 Nr. 3 GenDG bei genetischen Untersuchungen zur Klärung der Abstammung. *Bundesgesundheitsblatt*, 54, 1242–7.

German Bar Association. 2009. Stellungsnahme des Deutschen Anwaltsvereins durch den Ausschuss Ausländer- und Asylrecht zu §17 Abs. 8 des Entwurfes eines Gesetzes über genetisch Untersuchungen bei Menschen (Gendiagnostikgesetz – GenDG) – Bundesrats-Drucksache 633/08 vom 29.08.2008, [Online] Available at: http://anwaltverein.de/downloads/Stellungnahmen-09/SN0609.pdf [accessed 24 April 2014].

Gibbon, S., 2002. Re-examining geneticization: Family trees in breast cancer genetics. *Science as Culture*, 11(4), 429–57.

Gibson, M., 2001. The truth machine: Polygraphs, popular culture and the confessing body. *Social Semiotics*, 11(1), 61–73.

Gjertson, D.W., Brenner, C.H., Baur, M.P., Carracedo, A., Guidet, F., Luque, J.A., Lessig, R., Mayr, W.R., Pascali, V.L., Prinz, M., Schneider, P.M. and Morling, N., 2007. ISFG: Recommendations on biostatistics in paternity testing. *Forensic Science International: Genetics*, 1(3–4), 223–31.

Gomes-Cardoso, J., 2009. *DNA Testing for Family Reunification: Fortress Europe's Latest Tool to Exclude the Unwanted?* MA Thesis. Katholieke Universiteit Leuven, Leuven.

Görgülü v. *Germany* 2004. *Application no. 74969/01.*

Groenendijk, K., 2006. Family reunification as a right under community law. *European Journal of Migration and Law*, 3(2), 215–30.

Hahn, D., 2007. Umstrittener Gesetzesentwurf: Frankreich entdeckt Einwanderer-Gen. *die tageszeitung*, [Online 24 October] Available at: http://www.taz.de/Umstrittener-Gesetzesentwurf/!6496/ [accessed 24 April 2014].

Hammersley, M. and Atkinson, P., 2007. *Ethnography: Principles in Practice.* Third Edition. Oxford and New York, NY: Routledge.

Hautaniemi, P., 2004. *Pojat! Somalipoikien kiistanalainen nuoruus Suomessa.* Helsinki: Nuorisotutkimusverkosto.

Hautaniemi, P., 2007. Diasporic authenticity: Connecting genes and building families through DNA testing in Somali family reunification in finland. In: A. Kusow and S. Björk, eds. *From Mogadishu to Dixon: The Somali Diaspora in a Global Context.* Trenton, NJ: Red Sea Press, 119–234.

HaVL, 1999a. *Ulkomaalaisvirasto, Kuuleminen Eduskunnan hallintovaliokunnassa*, 26 October 1999.

HaVL, 1999b. *Pakolaisneuvonta, Kuuleminen Eduskunnan hallintovaliokunnassa*, 26 October 1999.

HaVL, 1999c. *Helsingin yliopisto, oikeuslääketieteen laitos, Jatkolausunto Eduskunnan hallintovaliokunnalle laiksi ulkomaalaislain muuttamisesta (HE 88/1999)*, 26 October 1999.

HaVL, 1999d. *Työministeriö, Lausunto Eduskunnan hallintovaliokunnalle laiksi ulkomaalaislain muuttamisesta (HE 88/1999)*, 26 October 1999.

HaVL, 1999e. *Sosiaali- ja terveysministeriö, Lausunto Eduskunnan hallintovaliokunnalle laiksi ulkomaalaislain muuttamisesta (HE 88/1999)*, 26 October 1999.

HE 88/1999. *Hallituksen esitys Eduskunnalle laiksi ulkomaalaislain muuttamisesta*, [Online] Available at: http://www.finlex.fi/fi/esitykset/he/1999/19990088 [accessed 24 April 2014].

HE 240/2009. *Hallituksen esitys Eduskunnalle laiksi ulkomaalaislain muuttamisesta*, [Online] Available at: http://www.finlex.fi/fi/esitykset/he/2009/20090240 [accessed 24 April 2014].

Heath, D., Rapp, R. and Taussig, K.-S., 2004. Genetic citizenship. In: D. Nugent and J. Vincent, eds. *A Companion to the Anthropology of Politics*. Malden, MA: Blackwell Publishers, 152–67.

Heinemann, T. and Lemke, T., 2013. Suspect families: DNA kinship testing in German immigration policy. *Sociology*, 47(4), 810–26.

Heinemann, T. and Lemke, T., 2014. Biological citizenship reconsidered: The use of DNA analysis by immigration authorities in Germany. *Science, Technology and Human Values*, 39(4), 488–510.

Heinemann, T., Lemke, T. and Prainsack, B., 2012. Risky profiles: Societal dimensions of forensic uses of DNA profiling technologies. *New Genetics and Society*, 31(3), 249–58.

Heinemann, T., Naue, U. and Tapaninen, A.-M., 2013. Verifying the family? A comparison of DNA analysis for family reunification in three European countries (Austria, Finland and Germany). *European Journal of Migration and Law*, 15(2), 183–202.

Helén, I., 2005. Genealogia kritiikkinä. *Sosiologia*, 42(2), 93–109.

Helén, I., 2014. Biological citizenship across the borders: Politics of DNA profiling for family reunification. *Distinktion: Scandinavian Journal of Social Theory*, epub ahead of print 9 June 2014. Doi: 10.1080/1600910X.2014.923321.

Helén, I. and Tapaninen, A.-M. 2013. Closer to the Truth: DNA Profiling for Family Reunification and the Rationales of Immigration Policy in Finland. *Nordic Journal of Migration Research*, 3(3), 153–61.

Hindmarsh, R. and Prainsack, B. (eds), 2010. *Genetic Suspects. Global Governance of Forensic DNA Profiling and Databasing*. Cambridge, UK and New York, NY: Cambridge University Press.

Holland, A., 2008. The modern family unit: Toward a more inclusive vision of the family in immigration law. *California Law Review*, 96(4), 1049–91.

Holland, E., 2011. Moving the virtual border to the cellular level: Mandatory DNA testing and the U.S. refugee family reunification program. *California Law Review*, 99(6), 1635–81.

Honahan, I., 2009. Reconsidering the claim to family reunification in migration. *Political Studies*, 57(4), 768–87.

Huber, B. (ed.), 2010. *Aufenthaltsgesetz*. München: Beck.

Huddleston, T., 2011. *Right to Family Reunion: The Dynamics Between EU Law and National Policy Change*. Brussels: Migration Policy Group.

Husserl, E., 1970. *The Crisis of European Science and Transcendental Phenomenology*. Evanston, IL: Northwestern University Press.

Huysmans, J., 2000. The European Union and the securitization of migration. *Journal of Common Market Studies*, 38(5), 751–77.

Iacono, W.G. and Lykken, D.T., 1997. The validity of the lie detector: Two surveys of scientific opinion. *Journal of Applied Psychology*, 82(3), 426–33.

International Federation for European Law, 2004. *Migration and Asylum Law and Policy in the European Union*. Cambridge, UK: Cambridge University Press.

Isin, E.F., Nyers, P., and Turner, B.S., 2008. *Citizenship Between Past and Future*. London: Routledge.

Jastram, K. and Newland, K., 2003. Family unity and refugee protection. In: E. Feller, V. Türk and F. Nicholson, eds. *Refugee Protection in International Law: UNHCR's Global Consultations on International Protection*. Cambridge, UK and New York, NY: Cambridge University Press, 555–603.

Jeffreys, A.J., Brookfield, J.F. and Semeonoff, R., 1985. Positive identification of an immigration test-case using human DNA fingerprints. *Nature*, 317(6040), 818–9.

Jeffreys, A.J., Wilson, V. and Thein, S.L., 1985a. Hypervariable 'minisatellite' regions in human DNA. *Nature*, 314(6006), 67–73.

Jeffreys, A.J., Wilson, V. and Thein, S.L., 1985b. Individual-specific 'fingerprints' of human DNA. *Nature*, 316(6023), 76–9.

John, A., 2003. *Family Reunification for Migrants and Refugees: A Forgotten Human Right?* MA Thesis. Human Rights Centre, University of Coimbra, [Online] Available at: http://www.fd.uc.pt/hrc/pdf/papers/arturojohn.pdf [accessed 24 April 2014].

Jones, A.G. and Ardren, W.R., 2003. Methods of parentage analysis in natural populations. *Molecular Ecology*, 12(10), 2511–23.

Jones, A.G., Small, C.M., Paczolt, K.A. and Ratterman, N.L., 2010. A practical guide to methods of parentage analysis. *Molecular Ecology Resources*, 10(1), 6–30.

Junge, A., Brinkmann, B., Fimmers, R. and Madea, B., 2006. Mutations or exclusion: An unusual case in paternity testing. *International Journal of Legal Medicine*, 120(6), 360–63.

Karlsson, A.O., Holmlund, G., Egeland, T., and Mostad, P., 2007. DNA-testing for immigration cases: The risk of erroneous conclusions. *Forensic Science International*, 172(2–3), 144–9.

Kautzor v. *Germany* 2012. *Application no 23338/09*.

Kern, B.-R. (ed.), 2012. *Gendiagnostikgesetz Kommentar*. München: Beck.

Kerr, A., 2003. Genetics and citizenship. *Society*, 40(6), 44–50.

Khosravi, S., 2009. Detention and deportation of asylum seekers in Sweden. *Race and Class*, 50(4), 30–56.

Koch, L., 2002. The government of genetic knowledge. In: S. Lundin and L. Akesson, eds. *Gene Technology and Economy: An Interdisciplinary Perspective*. Lund: Nordic Academic Publishers, 92–103.

Kofman, E., 2004. Family-related migration: A critical review of European Studies. *Journal of Ethnic and Migration Studies*, 30(2), 243–62.

Kokkarinen, H., (n.d.). *Pakolaisten perheenyhdistäminen – DNA-tutkimuksen käyttö perheenyhdistämistapauksissa*. Unpublished report, Pakolaistoimisto, Sosiaali- ja terveysministeriö.

Konrad, M., 2005. *Narrating the New Predictive Genetics: Ethics, Ethnography and Science*. Cambridge, UK: Cambridge University Press.

Kollek, R. and Lemke, T., 2008. *Der medizinische Blick in die Zukunft: Gesellschaftliche Implikationen prädiktiver Gentests*. Frankfurt am Main and New York, NY: Campus.

Kopelmann, L.M., 2004. Children III: Healthcare and research issues. In: S.G. Post, ed. *Encyclopedia of Bioethics*. Third Edition. New York, NY: Macmillan, 385–98.

Kraler, A., Hollomey, C., Hurich, C., König, A. and Muzak, G., 2013. *Family Reunification: A Barrier or Facilitator to Integration?*. Country Report Austria, [Online] Available at: http://research.icmpd.org/fileadmin/Research-Website/Project_material/Family_Reunification_EIF/Austria1.pdf [accessed 24 April 2014].

Kreienbrink, A. and Rühl, S., 2007. *Family Reunification in Germany: Small Scale Study IV in the Framework of the European Migration Network*. Nuremberg: Federal Office for Migration and Refugees.

Kruse, C., 2010. Producing absolute truth: *CSI* science as wishful thinking. *American Anthropologist*, 112(1), 79–91.

Kruse, C., 2012. Legal storytelling in pre-trial investigations: Arguing for a wider perspective on forensic evidence. *New Genetics and Society*, 31(3), 299–309.

La Spina, E., 2012. DNA testing for family reunification in Europe: An exceptional resource? *Migraciones Internacionales*, 6(3), 39–74.

Lakhani, S.M. and Timmermans, S., 2014. Biopolitical citizenship in the immigration adjudication process. *Social Problems*, 61(3), 360–79.

Leitzinger, A. (ed.), 2010. *Mansikkamaanvartijat: Muistelmia ulkomaalaishallinnosta eri vuosikymmeniltä*. 2010. Helsinki: Maahanmuuttovirasto. [Online] Available at: http://www.migri.fi/download/16289_mansikkamaan_vartijat.pdf [accessed 24 April 2014].

Lemke, T., 2004. Disposition and determinism – genetic diagnostics in risk society. *The Sociological Review*, 52(4), 550–66.

Lemke, T. and Wehling, P., 2009. Bürgerrechte durch Biologie? Kritische Anmerkungen zur Konjunktur des Begriffs 'biologische Bürgerschaft'. In: M.G. Weiss, ed. *Bios und Zoë. Die menschliche Natur im Zeitalter ihrer technischen Reproduzierbarkeit*. Frankfurt am Main: Suhrkamp, 72–107.

Lepola, O., 2000. *Ulkomaalaisesta suomenmaalaiseksi*. Helsinki: Finnish Literature Society.

Lippert, R.K. and Pyykkönen, M., 2012. Contesting family in Finnish and Canadian immigration and refugee policy. *Nordic Journal of Migration Research*, 2(1), 45–56.

Lippman, A., 1991. Prenatal genetic testing and screening: Constructing needs and reinforcing inequities. *American Journal of Law and Medicine*, 17(1–2), 15–50.

Lister, M., 2010. Immigration, association, and the family. *Law and Philosophy*, 2010/6, 717–45.

Ljungqvist, A. and Simpson, J.L., 1992. Medical examination for health of all athletes replacing the need for gender verification in international sports: The International Amateur Athletic Federation Plan. *Journal of the American Medical Association*, 267(6), 850–52.

Lynch, M., Cole, S.A. and McNally, R., 2008. *Truth Machine: The Contentious History of DNA Fingerprinting*. Chicago, IL: University of Chicago Press.

Maahanmuuttovirasto 2011. *Tilastokatsaus 2010* [Online] Available at: http://www.migri.fi/download/16636_Mamun_tilastokatsaus32010.pdf [accessed 24 April 2014].

Maahanmuuttovirasto, 2012. *Vuosikertomus 2011*, [Online] Available at: http://issuu.com/maahanmuuttovirasto/docs/vuosikertomus_2011 [accessed 24 April 2014].

Maahanmuuttovirasto, 2014. Tilastokatsaus, Maahanmuuttoyksikkö 2013, [Online] Available at: http://www.migri.fi/download/50381_Tilastokatsaus_Oleskeluluvat_2013_pdf.pdf [accessed 24 April 2014].

Machado, H. and Prainsack, B., 2012. *Tracing Technologies: Prisoners' Views in the Era of CSI*. Farnham: Ashgate.

Mansuet-Lupo, A., Henke, J., Henke, L., Blank, C., Ernsting, A., Kozlowski, P., Rouger, P. and van Huffel, V., 2009. A paternity case with three genetic incompatibilities between father and child due to maternal uniparental disomy 21 and a mutation at the Y chromosome. *Forensic Science International: Genetics*, 3(2), 141–3.

Marshall, T.H., 1950. *Citizenship and Social Class and Other Essays*. Cambridge, UK: Cambridge University Press.

M'charek, A., 2005. *The Human Genome Diversity Project: An Ethnography of Scientific Practice*. Cambridge, UK and New York, NY: Cambridge University Press.

M'charek, A., 2008. Silent witness, articulate collective: DNA evidence and the inference of visible traits. *Bioethics*, 22(9), 519–28.

McKie, R., 2009. Eureka moment that led to the discovery of DNA fingerprinting. *The Observer*, [Online 24 May] Available at: http://www.guardian.co.uk/science/2009/may/24/dna-fingerprinting-alec-jeffreys [accessed 24 April 2014].

Mill, J.S., 1983. *On Liberty*. Cambridge, UK: Cambridge University Press.

Mindle, G.B., 1989. Liberalism, privacy, and autonomy. *The Journal of Politics*, 51(3), 575–98.

Mol, A., 2002. *The Body Multiple: Ontology in Medical Practice*. Durham, NC: Duke University Press.

Morling, N., Allen, R.W., Carracedo, A., Geada, H., Guidet, F., Hallenberg, C., Martin, W., Mayr, W.R., Olaisen, B., Pascali, V.L. and Schneider, P.M., 2002. Paternity Testing Commission of the International Society of Forensic Genetics: Recommendations on genetic investigations in paternity cases. *Forensic Science International*, 129(3), 148–57.

Motion for a Resolution Regarding DNA Analysis as a Tool to Control Immigration, 2008. Entschließungsantrag der Abgeordneten Strache, Vilimsky und weiterer Abgeordneter betreffend DNA-Tests zur Immigrationskontrolle. 274/A(E) XXIV. Gesetzgebungsperiode. 10. Dezember 2008, [Online 10 December] Available at: http://www.parlament.gv.at/PAKT/VHG/XXIV/A/A_00274/fname_145824.pdf [accessed 24 April 2014].

Muller, B.J., 2004. (Dis)qualified bodies: Securitization, citizenship and 'identity management'. *Citizenship Studies*, 8(3), 279–94.

Müller, A., 2012. *Misuse of the Right to Family Reunification: Marriages of Convenience and False Declarations of Parenthood.* Nuremberg: Federal Office for Migration and Refugees.

Murdock, T.R., 2008. Whose child is this? Genetic analysis and family reunification immigration in France. *Vanderbilt Journal of Transnational Law*, 41(5), 1503–34.

Nayar, P.K., 2012. 'I Sing the Body Biometric'. *Economic and Political Weekly*, 47(32), 17–22.

Novas, C. and Rose, N., 2000. Genetic risk and the birth of the somatic individual. *Economy and Society*, 29(4), 485–513.

OM., 1999. *Oikeusministeriön lausunto sisäasiainministeriölle luonnoksesta hallituksen esityksestä ulkomaalaislain muuttamisesta*, 6 August 1999.

Opsahl, T., 1973. The convention on the right to respect for family life, particularly as regards the unity of the family and the protection of the rights of parents and guardians in the education of children. In: A.H. Robertson, ed. *Privacy and Human Rights*, Manchester: Manchester University Press, 182–255.

Parent, W.A., 1983. Privacy, morality, and the law. *Philosophy and Public Affairs,* 12(4), 269–88.

Petryna, A., 2002. *Life Exposed: Biological Citizens After Chernobyl.* Princeton, NJ: Princeton University Press.

Pöry, 2010. *Marriages of Convenience: A Comparative Study. Rules and Practices in Norway, Sweden, Germany, Denmark and the Netherlands. Commissioned by the Norwegian Directorate of Immigration (UDI).* Stavanger: Pöry Management Consulting.

Prainsack, B., Reardon, J., Hindmarsh, R., Gottweis, H., Naue, U. and Lunshof J.E., 2008. Personal genomes: Misdirected precaution. *Nature*, 456(7218), 34–5.

Prainsack, B., 2010. Key issues in DNA profiling and databasing: Implications for governance. In: R. Hindmarsh and B. Prainsack, eds. *Genetic Suspects: Global Governance of Forensic DNA Profiling and Databasing*. Cambridge, UK and New York, NY: Cambridge University Press, 15–39.

President's Commission for the Study of Ethical Problems in Medicine and Biomedical and Behavioral Research, 1982. *Making Health Care Decisions: A Report on the Ethical and Legal Implications of Informed Consent in the Patient-Practitioner Relationship.* Washington D.C.

Raman, S. and Tutton, R., 2010. Life, science, and biopower. *Science, Technology and Human Values*, 35(5), 711–34.

Reid, J.E., 1947. A revised questioning technique in lie-detection tests. *Journal of Criminal Law and Criminology*, 37(6), 542–7.

Reubi, D., 2010. Blood donors, development and modernisation: Configurations of biological sociality and citizenship in post-colonial Singapore. *Citizenship Studies*, 14(5), 473–93.

Rheinberger, H.-J., 1992. Experiment, difference, and writing: II. The laboratory production of transfer RNA. *Studies in History and Philosophy of Science*, 23A(3), 389–422.

Richards, M., 2010. DNA paternity testing, parentage and kinship: Reflections on some tendencies in the UK and in the USA. *L'Homme*, 21(2), 101–5.

Rose, N., 2007. *The Politics of Life Itself: Biomedicine, Power and Subjectivity in the Twenty-First Century*. Princeton, NJ and Oxford: Princeton University Press.

Rose, N., and Novas, C., 2005. Biological citizenship. In: S.J. Collier and A. Ong, eds. *Global assemblages: Technology, Politics, and Ethics as Anthropological Problems*. Malden, MA: Blackwell Publishers, 439–63.

Rouvroy, A., 2008. *Human Genes and Neoliberal Governance: A Foucauldian Critique*. London and New York, NY: Routledge.

Ruffer, G., 2011. Pushed beyond recognition? The liberality of family reunification policies in the EU. *Journal of Ethnic and Migration Studies*, 37(6), 935–51.

Schaffer, R., Kuczynski, K. and Skinner, D., 2008. Producing genetic knowledge and citizenship through the Internet: Mothers, pediatric genetics, and cybermedicine. *Sociology of Health and Illness*, 30(1), 145–59.

Schibel, Y., 2004. Transposing EU law on legal migration. *European Journal of Migration and Law*, 6(4), 395–404.

Schmidt, W., 2007. Ämter bei Ausländerkindern zu neugierig. *die tageszeitung*, 6 November.

SDK 378/1991. *Ulkomaalaislaki*, [Online] Available at: http://www.finlex.fi/fi/laki/alkup/1991/19910378 [accessed 24 April 2014].

SDK 731/1999. *Suomen perustuslaki*, [Online] Available at: http://www.finlex.fi/fi/laki/alkup/1999/19990731 [accessed 24 April 2014].

SDK 114/2000. *Laki ulkomaalaislain muuttamisesta*, [Online] Available at: http://www.finlex.fi/fi/laki/alkup/2000/20000114 [accessed 24 April 2014].

SDK 301/2004. *Ulkomaalaislaki*, [Online] Available at: http://www.finlex.fi/fi/laki/ajantasa/2004/20040301 [accessed 24 April 2014].

SDK 378/2005. *Laki oikeusgeneettisestä isyystutkimuksesta*, [Online] Available at: http://www.finlex.fi/fi/laki/ajantasa/2005/20050378 [accessed 24 April 2014].

SDK 755/2005. *Valtioneuvoston asetus oikeugeneettisestä isyystutkimuksesta*, [Online] Available at: http://www.finlex.fi/fi/laki/ajantasa/2005/20050755 [accessed 24 April 2014].

SDK 549/2010. *Laki ulkomaalaislain muuttamisesta*, [Online] Available at: http://www.finlex.fi/fi/laki/alkup/2010/20100549 [accessed 24 April 2014].

Segrave, K., 2004. *Lie Detectors. A Social History*. Jefferson, NC and London: McFarland.

Simpson, J.L., Ljungqvist, A., Ferguson-Smith, M.A., La Chapelle, A. de, Elsas, L.J., Ehrhardt, A.A., Genel, M., Ferris, E.A. and Carlson, A., 2000. Gender verification in the Olympics. *Journal of the American Medical Association*, 284(12), 1568–69.

Stacey, J., 1996. *In the Name of the Family: Rethinking Family Values in the Postmodern Age*. Boston, MA: Beacon Press.

Statistics Finland 2012. *Statistical Yearbook of Finland 2012*. Helsinki: Statistics Finland, [Online] Available at: http://pxweb2.stat.fi/sahkoiset_julkaisut/vuosikirja2012/pdf/julkaisu.pdf [accessed 24 April 2014].

Stenographic Records, 2009. *Stenographisches Protokoll, 40. Sitzung des Nationalrates der Republik Österreich, XXIV. Gesetzgebungsperiode. 21. Oktober 2009*, [Online] Available at: http://www.parlament.gv.at/PG/DE/XXIV/NRSITZ/NRSITZ_00040/fname_177083.pdf [accessed 24 April 2014].

Swarns, R.L., 2007. DNA tests offer immigrants hope or despair. *New York Times*, [Online 10 April] Available at: http://www.nytimes.com/2007/04/10/us/10dna.html [accessed 24 April 2014].

Swedish Migration Board, 2012. *For persons from Somalia with children under the age of 18*, [Online] Available at: http://www.migrationsverket.se/info/5515_en.html [accessed 24 April 2014].

Taitz, J., Weekers, E.M. and Mosca, D.T., 2002a. DNA and immigration: The ethical ramifications. *The Lancet*, 359, 794.

Taitz, J., Weekers, J.E.M. and Mosca, D.T., 2002b. The last resort: Exploring the use of DNA testing for family reunification. *Health and Human Rights*, 6(1), 20–32.

Teinonen, M. and Virtanen, T.J. (eds), 2000. *Ingrians and Neighbours: Focus on the Eastern Baltic Sea Region*. Helsinki: Finnish Literature Society.

Ten Have, H.A.M.J., 2004. Geneticization: The socio-cultural impact of gentechnology. In: G. Magill, ed. *Genetics and Ethics: An Interdisciplinary Study*. Saint Louis, MO: Saint Louis University Press, 80–100.

Ticktin, M., 2011. How biology travels: A humanitarian trip. *Body and Society*, 17(2–3), 139–58.

Tishkoff, S.A., Reed, F.A., Friedlaender, F.R., Ehret, C., Ranciaro, A., Froment, A., Hirbo, J.B., Awomoyi, A.A., Bodo, J.-M., Doumbo, O., Ibrahim, M., Juma, A.T., Kotze, M.J., Lema, G., Moore, J.H., Mortensen, H., Nyambo, T.B., Omar, S.A., Powell, K., Pretorius, G.S., Smith, M.W., Thera, M.A., Wambebe, C., Weber, J.L. and Williams, S.M., 2009. The genetic structure and history of Africans and African Americans. *Science*, 324(5930), 1035–44.

Toom, V. and Prainsack, B., 2010. The Prüm regime: Situated dis/empowerment in transnational DNA profile exchange. *British Journal of Criminology*, 50(6), 1117–35.

Turney, L., 2005. The incidental discovery of nonpaternity through genetic carrier screening: An exploration of lay attitudes. *Qualitative Health Research*, 15(5), 620–34.

Turney, L., 2006. Paternity testing and the biological determination of fatherhood. *Journal of Family Studies*, 12(1), 73–93.

Turney, L., 2011. The denial of paternity: Pregnancy as a risk to the 'pure relationship'. *Sociology*, 45(6), 1110–25.

Tutton, R., Hauskeller, C. and Sturdy, S., 2014. Suspect technologies: Forensic testing of asylum seekers at the UK border. *Racial and Ethnic Study*, 1–17.

Ulkomaalaisvirasto (UVI) 1996. DNA-testien käyttömahdollisuudet Suomessa asuvien Somalien sekä heidän Etiopiassa ja Keniassa olevien lähisukulaistensa perheenyhdistämisessä. Unpublished report 5.7.1996.

Ulkomaalaisvirasto (UVI), 2001. *Ulkomaalaisviraston vuosikertomus 2000*. Helsinki: Ulkomaalaisvirasto.

Ulkomaalaisvirasto (UVI), 2007. *Ulkomaalaisviraston vuosikertomus 2006*, [Online] Available at: http://www.migri.fi/download/16312_Vuosikertomus_2006.pdf?b21dae3f9ab4d088 [accessed 24 April 2014].

Ulkomaalaisvaltuutettu (UVT), 1996a. *Toimintakertomus 1995*. Helsinki: Ulkomaalaisvaltuutetun toimisto.

Ulkomaalaisvaltuutettu (UVT), 1996b. *DNA-testauksen käyttömahdollisuudet Suomessa asuvien somalien sekä heidän Etiopiassa ja Keniassa olevien lähiomaistensa perheenyhdistämisessä*, Unpublished report. 5 July 1996.

Ulkomaalaisvaltuutettu (UVT), 1997. *Toimintakertomus 1996*. Helsinki: Ulkomaalaisvaltuutetun toimisto.

UNHCR, 1999. Family protection issues. *Refugee Survey Quarterly*, 18(2), 102–10.

UNHCR, 2008. *UNHCR Note on DNA Testing to Establish Family Relationships in the Refugee Context*, [Online] Available at: http://www.unhcr.org/refworld/docid/48620c2d2.html [accessed 24 April 2014].

van der Ploeg, I., 1999. The illegal body: 'Eurodac' and the politics of biometric identification. *Ethics and Information Technology*, 1(4), 295–302.

Verband binationaler Familien und Partnerschaften, 2012. Stellungnahme des Verbandes binationaler Familien und Partnerschaften, iaf e.V. zu: Grünbuch der Kommission zum Recht auf Familienzusammenführung von in der Europäischen Union lebenden Drittstaatsangehörigen (Richtlinie 2003/86/E.G.) KOM(2011) 735 endgültig, [Online] Available at: http://www.verband-binationaler.de/fileadmin/user_upload/Bundesverband/Stellungnahme_Gruenbuch_28_02_2012.pdf [accessed 24 April 2014].

Villiers, J.D., 2010. Brave new world: The use and potential misuse of DNA technology in immigration law. *Boston College Third World Law Journal*, 30(2), 239–71.

Walsh, S.J. and Buckleton, J.S., 2005. DNA intelligence databases. In: J.S. Buckleton, C.M. Triggs and S.J. Walsh, eds. *Forensic DNA Evidence Interpretation*. Boca Raton, London, New York, NY, and Washington, D.C.: CRC Press, 439–69.

Walter, A., 2009. *Familienzusammenführung in Europa: Völkerrecht, Gemeinschaftsrecht, Nationales Recht*. Baden-Baden: Nomos.

Walzer, M., 1984. *Spheres of Justice*. New York, NY: Basic.

Warren, S. and Brandeis, L.D., 1890. The right to privacy. *Harvard Law Review*, 4(5), 193–220.

Wear, St., Engelhard, H.T. and Wildes, K.W., 1997. *Informed Consent: Patient Autonomy and Clinician Beneficence Within Health Care*. Washington D.C.: Georgetown University Press.

Wehling, P., 2011. Biology, citizenship and the government of biomedicine: Exploring the concept of biology citizenship. In: U. Bröckling, S. Krasmann, and T. Lemke, eds. *Governmentality: Current Issues and Future Challenges*. New York, NY: Routledge, 225–46.

Weigel, S., 2002. Genealogie und Genetik: Schnittstellen zwischen Biologie und Kulturgeschichte. Berlin: Akademie Verlag.

Weiss, M.G., 2011. Strange DNA: The rise of DNA analysis for family reunification and its ethical implications. *Genetics, Society and Politics Journal*, 7, 1–19.

Wellman, C.H, 2008. Immigration and freedom of association. *Ethics*, 119, 109–41.

Westin, A.F., 1994. Privacy and genetic information. In: M.S. Frankel and A.H Teich, eds. *The Genetic Frontier: Ethics, Law, and Policy*. Washington D.C.: American Association for the Advancement of Science, 53–76.

Yanow, D., 1996. *How Does a Policy Mean? Interpreting Policy and Organizational Actions*. Washington, D.C.: Georgetown University Press.

Yanow, D., 2000. *Conducting Interpretive Policy Analysis*. London: Sage.

Yin, R.K., 2014. *Case Study Research: Design and Methods*. Fifth Edition. Thousand Oaks, CA: Sage.

Zadok, E., Ben-Or, G. and Fisman, G., 2010. Forensic utilisation of voluntarily collected DNA samples: Law enforcement versus human rights. In: R. Hindmarsh and B. Prainsack, eds. *Genetic Suspects: Global Governance of Forensic DNA Profiling and Databasing*. Cambridge, UK and New York, NY: Cambridge University Press, 40–62.

Index

Page numbers in italic indicate the figure. Notes are referenced as 1n2 (page 1 note 2).

Aas, K.F. 73
adopted children 25, 26, 39, 42
Alien Law Package (Austria) 62, 70, 97
Aliens Act (Finland) 6, 37, 39, 40, 41–2, 43, 47, 48, 51–2, 96–7, 99, 100
Aliens Act (Germany) 16
Aliens Departments (Germany) 16, 19, 20, 24n12, 95, 96, 97, 107
Allen, A.L. 80, 85
Asylum Act (Austria) 56, 57–8, 59, 60, 61, 62, 63, 69, 71, 72, 77, 78, 96
asylum seekers 15, 51, 52, 101, 107
  Austria 56, 57–8, 68–9, 70–71, 72
  Finland 35, 36, 40
  Germany 23, 29, 97
Austria 4, 6, 29, 48, 55–6, 87, 95, 96, 102, 103–4
  Asylum Act 56, 57–8, 59, 60, 61, 62, 63, 69, 71, 72, 77, 78, 96
  asylum seekers 56, 57–8, 68–9, 70–71, 72
  children 61, 62, 72, 77, 87, 95
  DNA analysis 6, 10, 29, 56, 60, 61, 63–4, 65–8, 70, 71–2, 76–8, 97, 98, 107–8
  DNA profiling 55–6, 60, 63, 67, 71, 73
  documentary evidence 57, 60, 62–3, 68, 69, 70–72, 102, 103, 107–8
  embassies 57, 60, 61, 62, 63–4, 66, 67–8, 69, 96, 100
  Federal Asylum Agency 56, 60, 61, 62, 63–4, 65, 68, 71–2, 76, 96, 100, 102
  fraud 43, 69, 70–71, 99
  interviews 6, 10, 57–8, 60, 68–9, 100, 102
  laboratories 63, 64–8, 93, 99
  NGOs 57n3, 58, 59, 61, 64, 66, 71, 96, 98, 99
  recognised refugees 56–7, 58, 59, 61–2, 68, 70, 72, 77, 97, 98
  'regular' immigrants 56, 57, 58, 72, 96
  reuniting person 58–60, 61, 62, 63, 64, 66, 67–8, 69, 70, 72–3, 76–8, 98
  Settlement and Residence Act 56, 57, 58–9, 70, 72, 96, 97
  sponsors 56, 58–9, 96
  trustworthiness 5, 56, 61, 68–9, 70–71, 72–3, 75, 76, 102

biological citizenship 7–8, 10, 92, 94, 104–7
biological family 10, 30, 101, 103, 107
  Germany 10, 17, 25–6, 31, 87
biological relationship 2–3, 6, 30, 49, 75, 76, 80, 84–5, 89–91, 92, 106
  Austria 99, 100, 102
  Finland 39, 48, 50, 100, 102, 103
  Germany 13, 14, 17, 19–20, 25–8, 29, 31, 86–7, 99–100, 102
biometric identification 49, 73, 101
biotechnological testing 50, 52, 101

Charter of Fundamental Rights (EU) 5, 94
child trafficking 2, 79, 82, 83, 92
children 2–3, 49, 79, 83–4, 87, 90, 92, 103
  Austria 61, 62, 72, 77, 87, 95
  Finland 37, 38, 39, 41–2, 45, 47, 51, 95, 103
  Germany 15, 16, 17–18, 19–20, 25–6, 87, 95
  interviews 82–3, 84
citizens (Germany) 14, 17, 22n10, 23, 26, 27, 86, 97

Convention on the Rights of the Child (CRC) 83–4, 87
Council Directive 2003/86/EC 5–6, 14–15, 41, 82, 94–5, 100
criminal activity 23, 29, 70–71, 73, 74, 86, 108

data storage 3, 23, 67, 79, 84–5, 86, 108
DNA analysis 1–4, 5–6, 29–30, 49, 73, 79–80, 84–7, 91–2, 93–4, 97–100, 101–4, 105–7, 108–9
  Austria 6, 10, 29, 56, 60, 61, 63–4, 65–8, 70, 71–2, 76–8, 97, 98, 107–8
  European Union 24, 51
  Finland 6, 10, 33–4, 37–41, 42–3, *44*, 45–7, 48, 50–52, 96–7, 98–9
  Germany 6, 10, 13–14, 16, 17, 18–20, 22, 29, 30–31, 86, 97, 98, 108
  immigration policies 1–2, 3, 4, 9, 49, 106
  lie detectors 6, 10, 29, 56, 73, 75, 76, 77
DNA profiling 9, 21, 22, 51, 73, 108
  Austria 55–6, 60, 63, 67, 71, 73
  Germany 14, 23, 108
DNA testing, *see* DNA analysis
documentary evidence 5, 16, 70, 100, 103, 104
  Austria 57, 60, 62–3, 68, 69, 70–72, 102, 103, 107–8
  Finland 36, 37, 41–2, 45, 47, 49, 50, 100, 102, 103
  Germany 16–17, 20, 21, 24–5, 26, 27, 99–100, 103
Dublin II 55n1, 57

embassies 49
  Austria 57, 60, 61, 62, 63–4, 66, 67–8, 69, 96, 100
  Germany 16, 17, 20, 26, 95, 96, 107
European Convention on Human Rights (ECHR) 2, 4, 5, 91
European Court of Human Rights 4–5, 25
European Migration Network 6, 37n5, 51
European Union 4–6, 24, 51, 93, 94–5, 101

family 2, 3, 5, 28–30, 31, 61, 88–91, 94–5, 103; *see also* biological family; social family
family reunification 1–3, 4–7, 29–30, 78, 79–80, 107–9
Federal Asylum Agency (Austria) 56, 60, 61, 62, 63–4, 65, 68, 71–2, 76, 96, 100, 102
Federal Foreign Office (Germany) 16–17, 18, 19, 24
Finkler, K. 28
Finland 4, 6–7, 34–5, 95, 100–101, 102–4
  Aliens Act 6, 40, 43, 47, 48, 51–2, 99, 100
  asylum seekers 35, 36, 40
  children 37, 38, 39, 41–2, 45, 47, 51, 95, 103
  DNA analysis 6, 10, 33–4, 37–41, 42–3, *44*, 45–7, 48, 50–52, 96–7, 98–9
  documentary evidence 36, 37, 41–2, 45, 47, 49, 50, 100, 102, 103
  fraud 10, 44–7, 48, 50, 51, 52, 99
  human rights 10, 37–8, 40–41, 44, 50–51
  immigration authorities 36, 39, 45, 46, 47, 48, 50
  immigration management 33, 34, 35, 39, 49
  immigration policies 33, 48, 50, 52
  interviews 42, 45, 47–8, 52, 100, 102
  laboratories 38, 42, 93, 98–9
  recognised refugees 96, 97
  social family 39, 41–2, 50, 84, 100
  Somalis 35–6, 37–9, 43, 45, 49, 103, 106
  sponsors 42, 96, 103
  voluntary testing 39, 43
Finnish Immigration Service (Migri) 33, 42–3, 49, 91–2, 96, 99n3, 100, 101, 102
foster children 26, 42, 47, 52, 95, 103
France 13, 27, 90, 91
fraud 6, 10, 24, 51, 101
  Austria 43, 69, 70–71, 99
  Finland 10, 44–7, 48, 50, 51, 52, 99
  Germany 16, 24, 25, 30
Fried, C. 81

genetic citizenship, *see* biological citizenship
Genetic Diagnostics Act (Germany) 22–3, 86, 97
'genetic fingerprint' 1
genetic privacy 85
geneticisation 14, 16, 20–21, 28–9, 50
Germany 4, 6, 9–10, 15–16, 84, 86–7, 95, 102, 107, 108
  Aliens Departments 16, 19, 20, 24n12, 95, 96, 97, 107
  asylum seekers 23, 29, 97
  biological family 10, 17, 25–6, 31, 87
  citizens 14, 17, 22n10, 23, 26, 27, 86, 97
  DNA analysis 6, 10, 13–14, 16, 17, 18–20, 22, 29, 30–31, 86, 97, 98, 108
  DNA profiling 14, 23, 108
  documentary evidence 16–17, 20, 21, 24–5, 26, 27, 99–100, 103
  embassies 16, 17, 20, 26, 95, 96, 107
  Federal Foreign Office 16–17, 18, 19, 24
  fraud 16, 24, 25, 30
  immigration authorities 6, 10, 13, 18, 19, 20, 24–5, 27, 29, 31
  laboratories 19, 20–21, 27, 93, 98
  parental testing 9–10, 13–14, 17–18, 19, 20–21, 24, 25–6, 27, 30, 48, 97, 99–100
  recognised refugees 17, 97
  Residence Act 14–15, 18
  social family 10, 14, 25, 26, 27, 31, 87
  sponsors 15, 16, 95, 98
  voluntary testing 18, 20, 21, 23, 24, 30
Grounded Theory 8–9, 34–5

human rights 92, 107
  Finland 10, 37–8, 40–41, 44, 50–51
human trafficking 45, 92
  children 2, 79, 82, 83, 92

immigration authorities 2, 51, 82, 101, 103
  Finland 36, 39, 45, 46, 47, 48, 50
  Germany 6, 10, 13, 18, 19, 20, 24–5, 27, 29, 31
immigration management 6, 7, 8–9, 49, 55, 102, 106, 107
  Finland 33, 34, 35, 39, 49
immigration policies 5, 47, 50, 51, 52, 91, 94, 103, 107
  DNA analysis 1–2, 3, 4, 9, 49, 106
  Finland 33, 48, 50, 52
informational self-determination 22, 23, 31, 86
informed consent 3, 10, 40, 79, 85–6, 87, 92, 108
Ingrians (Finland) 35, 36
interviews 16, 81–3, 84, 91–2
  Austria 6, 10, 57–8, 60, 68–9, 100, 102
  Finland 42, 45, 47–8, 52, 100, 102

Jeffreys, A.J. 1

laboratories
  Austria 63, 64–8, 93, 99
  Finland 38, 42, 93, 98–9
  Germany 19, 20–21, 27, 93, 98
legal documents, *see* documentary evidence
liberty 80–81, 92
lie detectors 73, 74–6
  DNA analysis 6, 10, 29, 56, 73, 75, 76, 77
Lippman, A. 28

marriages of convenience 6, 51, 82
Migri, *see* Finnish Immigration Service
mistrust 23, 24, 30, 31, 47, 56, 70–71, 72, 94
Murdock, T.R. 27, 83, 91

NGOs (Austria) 57n3, 58, 59, 61, 64, 66, 71, 96, 98, 99
Novas, C., *see* Rose, N. and Novas, C.

official documents, *see* documentary evidence

Parent, W.A. 81
parental testing 2–3, 8, 29n17, 44, 55, 87
  Germany 9–10, 13–14, 17–18, 19, 20–21, 24, 25–6, 27, 30, 48, 97, 99–100

payment policies 20n9, 43n11, 66, 98, 99
polygraph 74–5
privacy protection 3, 31, 48, 80–81, 92
privacy, right to 3, 79, 80–81, 82, 91

recognised refugees
  Austria 56–7, 58, 59, 61–2, 68, 70, 72, 77, 97, 98
  Finland 96, 97
  Germany 17, 97
Refugee Advice Centre (Finland) 40
'regular' immigrants (Austria) 56, 57, 58, 72, 96
Residence Act (Germany) 14–15, 18
reuniting persons (Austria) 58–60, 61, 62, 63, 64, 66, 67–8, 69, 70, 72–3, 76–8, 98
right not to know 3, 10, 48–9, 86–7
Rose, N. and Novas, C. 104–5, 106

same sex partnerships 15, 25, 41, 61, 89, 95
sample destruction 23, 40, 85, 92, 108
sample storage 67n12, 79, 84–5, 86, 108
Settlement and Residence Act (Austria) 56, 57, 58–9, 70, 72, 96, 97
sham marriages, *see* marriages of convenience
social family 3, 28, 84, 87, 89–90, 91, 92
  Finland 39, 41–2, 50, 84, 100
  Germany 10, 14, 25, 26, 27, 31, 87
Somalis 27, 49, 69, 102n4
  Finland 35–6, 37–9, 43, 45, 49, 103, 106
sponsors 58, 59
  Austria 56, 58–9, 96
  Finland 42, 96, 103
  Germany 15, 16, 95, 98
storage, *see* data storage; sample storage
suspicion 14, 23, 45–6, 47, 48, 51, 101, 103
Sweden 49

Treaty on the Functioning of the European Union (TFEU) 5
trustworthiness 5–6, 23–4, 47, 75, 101, 103–4
  Austria 5, 56, 61, 68–9, 70–71, 72–3, 75, 76, 102
truth machines 10, 29, 50, 73–5; *see also* lie detectors

UN High Commissioner for Refugees (UNHCR) 59, 61, 89
Universal Declaration of Human Rights (UDHR) 2, 4, 108

voluntary testing 3, 13–14, 85–6, 92, 108
  Finland 39, 43
  Germany 18, 20, 21, 23, 24, 30

RESEARCH IN MIGRATION AND ETHNIC RELATIONS SERIES

Full series list

Intercultural Education in the European Context
Theories, Experiences, Challenges
Edited by Marco Catarci and Massimiliano Fiorucci

Diasporas and Homeland Conflicts
A Comparative Perspective
Bahar Baser

Employers, Agencies and Immigration
Paying for Care
Edited by Anna Triandafyllidou and Sabrina Marchetti

Inside Immigration Law
Migration Management and Policy Application in Germany
Tobias G. Eule

European Immigration
A Sourcebook
Edited by Anna Triandafyllidou and Ruby Gropas

Ethnic Diversity and Social Cohesion
Immigration, Ethnic Fractionalization and Potentials for Civic Action
Merlin Schaeffer

Irregular Migrant Domestic Workers in Europe
Who Cares?
Edited by Anna Triandafyllidou

Migrants and Cities
The Accommodation of Migrant Organizations in Europe
Margit Fauser

The Bosnian Diaspora
Integration in Transnational Communities
Edited by Marko Valenta and Sabrina P. Ramet

Negotiating National Identities
Between Globalization, the Past and 'the Other'
Christian Karner

Security, Insecurity and Migration in Europe
Edited by Gabriella Lazaridis

Media in Motion
Cultural Complexity and Migration in the Nordic Region
Edited by Elisabeth Eide and Kaarina Nikunen

Managing Ethnic Diversity
Meanings and Practices from an International Perspective
Edited by Reza Hasmath

Muslim Diaspora in the West
Negotiating Gender, Home and Belonging
Edited by Haideh Moghissi and Halleh Ghorashi

Inclusion and Exclusion of Young Adult Migrants in Europe
Barriers and Bridges
Edited by Katrine Fangen, Kirsten Fossan and Ferdinand Andreas Mohn

Irregular Migration in Europe
Myths and Realities
Edited by Anna Triandafyllidou

Paradoxes of Cultural Recognition
Edited by Sharam Alghasi, Thomas Hylland Eriksen and Halleh Ghorashi

Labour Migration from Turkey to Western Europe, 1960–1974
A Multidisciplinary Analysis
Ahmet Akgündüz

The African Diaspora in the United States and Europe
The Ghanaian Experience
John A. Arthur

Minority Rights Protection in International Law
The Roma of Europe
Helen O'Nions

Diversity Management and Discrimination
Immigrants and Ethnic Minorities in the EU
John Wrench

Cities and Labour Immigration
Comparing Policy Responses in Amsterdam, Paris, Rome and Tel Aviv
Michael Alexander

Immigrant Women and Feminism in Italy
Wendy Pojmann

Moving Lives
Narratives of Nation and Migration among Europeans in Post-War Britain
Kathy Burrell

Migration, Regional Integration and Human Security
The Formation and Maintenance of Transnational Spaces
Edited by Harald Kleinschmidt

Globalizing Migration Regimes
New Challenges to Transnational Cooperation
Edited by Kristof Tamas and Joakim Palme

Migration and its Enemies
Global Capital, Migrant Labour and the Nation-State
Robin Cohen

East to West Migration
Russian Migrants in Western Europe
Helen Kopnina

International Migration Research
Constructions, Omissions and the Promises of Interdisciplinarity
Edited by Michael Bommes and Ewa Morawska

Transnational Social Spaces
Agents, Networks and Institutions
Edited by Thomas Faist and Eyüp Özveren

Nationalism and Exclusion of Migrants
Cross-National Comparisons
Edited by Mérove Gijsberts, Louk Hagendoorn and Peer Scheepers

Citizenship in European Cities
Immigrants, Local Politics and Integration Policies
Edited by Rinus Penninx, Karen Kraal, Marco Martiniello and Steven Vertovec

International Migration in the New Millennium
Global Movement and Settlement
Edited by Danièle Joly

Governance in Multicultural Societies
Edited by John Rex and Gurharpal Singh

Identity and Integration
Migrants in Western Europe
Edited by Rosemarie Sackmann, Bernhard Peters and Thomas Faist

European Encounters
Migrants, Migration and European Societies Since 1945
Edited by Rainer Ohliger, Karen Schönwälder and Triadafilos Triadafilopoulos

Illegal Immigrants and Developments in Employment in the Labour Markets of the EU
Jan Hjarnø

Globalizing Chinese Migration
Trends in Europe and Asia
Edited by Pál Nyíri and Igor Saveliev

Integration and Resistance
The Relation of Social Organisations, Global Capital, Governments and International Immigration in Spain and Portugal
Ricard Morén-Alegret

Ethnicity and Nationalism in Italian Politics
Inventing the *Padania*: Lega Nord and the Northern Question
Margarita Gómez-Reino Cachafeiro

Poland's Post-War Dynamic of Migration
Krystyna Iglicka

South Africa's Racial Past
The History and Historiography of Racism, Segregation, and Apartheid
Paul Maylam

Patriarchal Structures and Ethnicity in the Italian Community in Britain
Azadeh Medaglia

Arguing and Justifying
Assessing the Convention Refugees' Choice of Moment, Motive and Host Country
Robert F. Barsky

Comparative Perspectives on Racism
Edited by Jessika ter Wal and Maykel Verkuyten

Transitions
Russians, Ethiopians, and Bedouins in Israel's Negev Desert
Edited by Richard Isralowitz and Jonathan Friedlander

European Nations and Nationalism
Theoretical and Historical Perspectives
Edited by Louk Hagendoorn, György Csepeli, Henk Dekker and Russell Farnen

New Chinese Migrants in Europe
The Case of the Chinese Community in Hungary
Pál Nyíri

Education and Racism
A Cross National Inventory of Positive Effects of Education on Ethnic Tolerance
Edited by Louk Hagendoorn and Shervin Nekuee

Long-Distance Nationalism
Diasporas, Homelands and Identities
Zlatko Skrbiš

Multicultural Policies and the State
A Comparison of Two European Societies
Edited by Marco Martiniello

Turks in European Cities
Housing and Urban Segregation
Edited by Sule Özüekren and Ronald Van Kempen

For Product Safety Concerns and Information please contact our EU
representative GPSR@taylorandfrancis.com
Taylor & Francis Verlag GmbH, Kaufingerstraße 24, 80331 München, Germany

www.ingramcontent.com/pod-product-compliance
Lightning Source LLC
LaVergne TN
LVHW020000031026
849161LV00021B/726
* 9 7 8 1 4 7 2 4 2 4 2 4 2 *